DO
WHALES
GET THE
BENDS
?

Also by Tony Rice

BRITISH OCEANOGRAPHIC VESSELS
1800–1950
•
DECOMMISSIONING THE BRENT SPAR
(with Paula Owen)
•
VOYAGES OF DISCOVERY
•
DEEP OCEANS

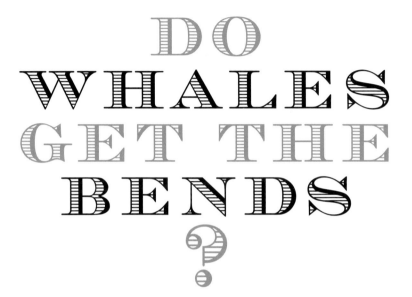

DO WHALES GET THE BENDS?

Answers to 118 Fascinating Questions about the Sea

TONY RICE

ADLARD COLES NAUTICAL
LONDON

Published by Adlard Coles Nautical
an imprint of A & C Black Publishers Ltd
36 Soho Square, London W1D 3QY
www.adlardcoles.com

First edition published 2010

ISBN 978-1-4081-1326-4

A CIP catalogue record for this book is available from the British Library.

This book is produced using paper that is made from wood grown in managed, sustainable forests.
It is natural, renewable and recyclable. The logging and manufacturing processes conform
to the environmental regulations of the country of origin.

Typeset in Agfa Rotis Semisans 10/13.5pt by Palimpsest Book Production Ltd,
Grangemouth, Stirlingshire
Printed and bound in Spain by GraphyCems

Inside front cover: Up, up and away! An Atlantic flying fish, *Cheilopogon melanurus,* takes to the
air, leaving its take-off trail behind it

Contents

INTRODUCTION   **WHAT'S IT ALL ABOUT?**   ix

| SECTION 1 | **OCEAN FACTS AND FIGURES** | 1 |

1	How big are the oceans?	1
2	Where did all the water come from?	2
3	How old are the oceans?	2
4	How salty are the oceans?	5
5	Why is the sea salty?	6
6	How deep are the deepest parts of the oceans?	8
7	How does pressure change with depth in the sea?	9
8	Why is the sea blue?	9
9	Why do icebergs float?	10
10	Why are icebergs white?	11
11	Why are icebergs in the Arctic and Antarctic so different?	12
12	What is sea ice?	14
13	What are the Continental Shelves?	14
14	What are the Continental Slopes?	15
15	What are the Abyssal Plains?	16
16	What are Deep Sea Trenches?	16
17	Is the bottom of the deep sea flat?	17
18	What are Hydrothermal Vents?	20
19	What is Coriolis Force?	24
20	What causes ocean currents?	27
21	Are there deep currents in the ocean?	29
22	What are Tectonic Plates?	31

| SECTION 2 | **LIFE IN THE OCEANS** | 33 |

GENERAL

23	Are seaweeds plants?	33
24	How deep do seaweeds grow in the ocean?	34
25	Why are seaweeds mainly brown?	35

26	What is plankton?	36
27	What is phytoplankton?	36
28	Why are some parts of the ocean more productive than others?	38
29	How deep in the oceans do animals live?	45
30	Why aren't deep sea animals crushed by the pressure?	45
31	How many different types of animals live in the ocean?	47
32	What do animals in the deep ocean feed on?	51
33	What are jellyfish?	53
34	What is the Portuguese Man-o'-war?	55
35	Does anything eat jellyfish?	56
36	Are there any marine insects?	57
37	What is bioluminescence?	57

SHELLFISH

38	What are shellfish?	58
39	Why does eating shellfish make some people very ill?	58
40	What are molluscs?	59
41	How big are giant squid?	60
42	What is the colossal squid?	62
43	What are crustaceans?	63
44	What is a lobster?	64
45	What is scampi?	68
46	What is the difference between a shrimp and a prawn?	69
47	What is or are krill?	71

FISH etc.

48	How many different types of fish live in the oceans?	72
49	Are all fish species cold blooded?	74
50	Which is the largest fish in the oceans?	75
51	How do flying fish fly?	75
52	Why do flying fish fly?	76
53	Do fish drink water?	77

BIRDS

54	Why do seabirds follow ships?	80
55	How many different seabirds are there?	81
56	What are seagulls?	83
57	How many penguin species are there?	83
58	How deep do penguins dive?	84
59	Do all penguins live in the Antarctic?	84
60	Where does the name penguin come from?	85
61	Do penguins have feathers or fur?	85

WHALES AND DOLPHINS

62	How many whale and dolphin species are there in the oceans?	86
63	How deep do whales dive?	87
64	Do all whales and dolphins come to the surface?	87
65	Do whales have to breathe out like scuba divers when they surface?	88
66	Do whales get the bends?	88
67	What's in a whale's blow?	90
68	Why do the blows of different whales look different?	91
69	Why is the Sperm whale called a Sperm whale?	91
70	What is the Sperm whale's spermaceti for?	93
71	Are dolphins whales?	94
72	What is the difference between a dolphin and a porpoise?	94
73	Why are Right whales called Right whales?	95
74	What are rorqual whales?	96
75	What is whalebone or baleen?	96
76	How do whales sing?	97
77	Why do whales sing?	98

SECTION 3   SHIPS AND SAILORS   99

78	What is the difference between a ship and a boat?	99
79	What is a fathom?	100
80	What is a nautical mile?	100
81	What are lines of latitude and longitude?	100
82	What is the difference between true north and magnetic north?	103
83	How does a gyrocompass work?	105
84	Why do Greenland and Antarctica look so huge on most world maps?	106
85	Why is a knot called a knot?	109
86	Why port and starboard?	109
87	What are all those marks on a ship's hull for?	110
88	What does a ship's tonnage mean?	112
89	Why do many ships have bulbous bows?	112
90	How far can you see from a ship at sea? (The simple version for most of us)	113
91	How far can you see from a ship at sea? (The nerd's version)	114
92	How far off can you see land from a ship at sea?	117

SECTION 4	WIND, WAVES AND WEATHER	118
93	How does the wind cause waves?	118
94	What is the Beaufort wind scale?	119
95	What does the Shipping Forecast mean?	121
96	Where on earth is Utseera?	122
97	Why are the Trade Winds called the Trade Winds?	123
98	What causes the Trade Winds?	124
99	How deep are the effects of waves felt?	127
100	Why do waves break when they reach the shore?	128
101	How big are the biggest oceanic waves recorded?	129
102	What are rogue waves?	130
103	What are tsunamis?	132

SECTION 5	TIDES	134
104	What are tides?	134
105	What causes the tides?	135
106	What are spring and neap tides?	140
107	Why are adjacent tides often of different heights?	142
108	Why do tides get later each day?	144
109	Where are the biggest tides in the world?	146
110	Why are there no tides in the Mediterranean?	146
111	What are tidal bores?	147

SECTION 6	THE OCEANS AND GLOBAL WARMING	149
112	Are the oceans getting warmer?	149
113	Is the sea level rising?	151
114	Do the oceans absorb carbon dioxide?	152

SECTION 7	MAN AND THE SEA	154
115	How deep have men been in the oceans?	154
116	Why do scuba divers have to breathe out when they return to the surface?	155
117	Does drinking seawater drive you mad?	155
118	Who owns the oceans?	158
INDEX		163

What's it all about?

The idea for this book grew from my experiences, since 2004, as a guest lecturer on cruise ships. My talks are broadly about the oceans and what makes them tick, based on a lifetime's career as a research marine biologist. Fortunately, this topic is relevant to any cruise and any ocean. So this fascinating new career has taken my wife and me across all the major oceans of the world and all latitudes from Greenland in the north to the Antarctic Peninsula in the south. Of course, we have enjoyed the exotic port calls enormously, from Saint Helena to Sydney and Bora Bora to Boston. But for us the sea passages have been even more fascinating. For me it has been like sailing through an oceanographic textbook, seeing for the first time phenomena and marine creatures that I had previously only read about.

Despite spending my professional life studying the oceans, my work had been confined largely to the Atlantic and particularly the north-eastern bit. So the opportunity to sail through the central Pacific during an El Niño year and to follow the track of Charles Darwin's famous voyage in HMS *Beagle* around South America, let alone the privilege of seeing dozens of species of whales, dolphins, sea birds and other marine creatures, has resurrected in me a boyish enthusiasm for all things maritime. It is this enthusiasm that I now want to share with you.

My main inspiration for writing the book stems from those generous individuals who have been kind enough to say that they have enjoyed my talks on marine life and maritime matters generally. Even more hearteningly, many of them have said that their increased interest in over-the-side matters has given the cruise an extra dimension. They are usually the ones who, like my wife and I, tramp the decks for hours on end hoping for a glimpse of a whale, a turtle, a flying fish or an exotic bird. Or if nothing like this appears, then some other interesting or baffling feature of the ocean itself. For example, I've often been asked 'Why is the sea that colour, and what causes those stripes on the surface?'. Fortunately, I knew the answers to those questions and most of the others I've been asked. But there have been some questions that I have had to admit completely stumped me. Finding out the answers to these, and putting this book together, has been a fascinating journey for me. I hope you enjoy reading it as much as I have enjoyed writing it.

The average length of an individual section or 'answer' is about 500 words; easily read over a cup of coffee, which was my intention. A few are much shorter, some as little as 100 words or less, while others, regrettably, are much longer because of the complicated nature of the questions. However, whether they are long or short, and despite lots of cross references, I have tried to make them stand-alone in the sense that you should be able to read, and understand, any of them without needing to read one of the others. But this carries with it a down side. To make the answers self-sufficient I've had to repeat one or two bits, sometimes several times. So if you read the book straight through you may, for example, get a little fed up with reading several times how the process of photosynthesis uses the energy of sunlight to build complex chemicals from simple molecules like carbon dioxide and water. I am sorry about this, but it is an inescapable consequence of this style of book and, I think, a small price to pay for the convenience of being able to dip into it.

Finally, since the book deals with the biggest environment on earth, it is not altogether surprising that it doesn't answer all the questions I've been asked over the past five years or so. If your pet question isn't included in this book, maybe it will be in the next one!

TONY RICE

Ocean Facts and Figures

1 How big are the oceans?

The world's oceans cover just about 71% of the surface area of the earth. With a total area of about 360 million square km[1] (140 million square miles) and an average depth of about 3.8 km (2.4 miles), the volume of the oceans is about 1,370 million cubic km or 330 million cubic miles. This represents about 97% of all the water on earth, compared with about 2% locked up (at the moment) in polar ice caps and glaciers, and the rest held in rivers, lakes and underground aquifers. Only about 0.0001% of the earth's water is held in the atmosphere at any time, though this is being continually removed by precipitation and replaced by evaporation.

The world human population in 2008 was about 6.7 billion. In the unlikely event that we needed to divide the ocean up between us, every man, woman and child alive today would get about one fifth of a cubic kilometre, that is some 200 million cubic metres. This is a somewhat meaningless figure, except that it illustrates just how vast the oceans are and suggests that we ought to be able to learn to manage them fairly efficiently. At the same time, with most estimates of population growth indicating that there will be more than 9 billion humans on earth by the year 2050, it is a sobering thought that by this time each person's share would have fallen by some 50 million cubic metres!

The oceans represent by far the biggest habitable environment on earth, with a volume estimated at about 160 times that of all other environments (earth/soil, air and freshwater) added together. Despite their huge size, however, the oceans make up less than 1/800th of the total volume of our planet (see also Q6).

---

1   About one million square kilometres for every day of the year! That's the way I remember it.

2 Where did all the water come from?

Nobody knows for certain, but wherever it came from, experts agree that it has been with us for a very long time.

The earth is estimated to be about 4.5 billion years old, with the oldest known rocks being about 3.7 billion years old. Some of these rocks are sedimentary, that is made up of material settling out from overlying water, so there must have been large bodies of liquid water on earth at least as early as this. Some of this water may have reached the earth in its early life associated with extraterrestrial dust particles but it is thought more likely that the water was part of the original complex mix of particles surrounding the sun in the very early days of the solar system and from which the earth and the other planets were eventually formed. So the water was probably part of the earth from its very beginning, initially locked up in some of its first rocks.

Eventually, but still fairly early on in the earth's history, the water would have been released from the rocks by volcanic activity and would have formed part of the hot atmosphere as water vapour or steam. Finally, when the atmosphere cooled a bit, the water vapour condensed and fell to earth as a warm rain and filled depressions in the earth's surface to form the first seas. So there was probably a separation between land areas and ocean areas on the earth's surface almost 4 billion years ago, though we have no idea what the relative proportions of land and water were at that time. Nor do we know what shape the oceans had, except that they were almost certainly nothing like the ones we see today (see Q3).

3 How old are the oceans?

As we saw in Q2, the earth has probably had liquid oceans for well over 3.5 billion of the 4.5 billion years it has been in existence. But the oceans as we see them today are much younger, resulting from the break-up of the super-continent *Pangea*, which began about 200 million years ago. In fact, it seems that this length of time is about average for the lifetime of individual ocean basins, after which they are replaced by new ones. So the oceans as we see them today are simply the most recent of a series of between 15 and 20 different sets of oceans that have existed since the first ones formed almost 4 billion years ago. We have no idea what shape most of these early oceans had, but immediately before the present system started to form, the situation appears to have been fairly simple.

At that time, most of the present day land masses were all joined together in one huge lump, *Pangea*, and occupied about half of the planet's surface. The other half was covered in a single vast ocean that palaeo-oceanographers call *Panthalassa*. Then, driven by what we now know as sea floor spreading or plate tectonics, *Pangea* began to break up around 170 million years ago. First, it split into two major parts, one that eventually produced most of what we call Europe and Asia, and the other, called by geologists *Gondwanaland*, consisting of what later became North and South America, Africa, India, Australia and Antarctica. The gap between these two land masses as they parted became a huge new ocean, *Tethys*, which gradually became obliterated as *Gondwanaland* broke up and its various pieces started to move towards their present positions (see Fig 1).

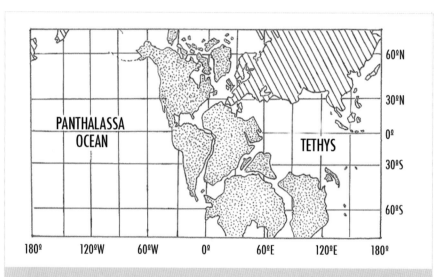

Figure 1
Distribution of the earth's land masses and oceans about 170 million years ago. By this time *Pangea* had already started to break up, with the future *Eurasia* (cross hatched) separating from *Gondwanaland* (stippled). The gap between them formed the new ocean, *Tethys*, eventually to become the Mediterranean and Indian Ocean. The ancient coastlines would not have corresponded to the modern ones shown on this fig and on Figs 2 and 3.

By about 100 million years ago (see Fig 2) Africa and the Americas started to separate both from one another, resulting in the beginnings of the Atlantic, and from Antarctica to form the Southern Ocean. At the same time, what is now Australia moved away from modern Antarctica to block the eastern side of *Tethys* and form the beginnings of the Indian Ocean. Finally, a triangular mass that eventually formed the Indian subcontinent moved through the new Indian Ocean, eventually to crash into *Eurasia* and force up the Himalayas.

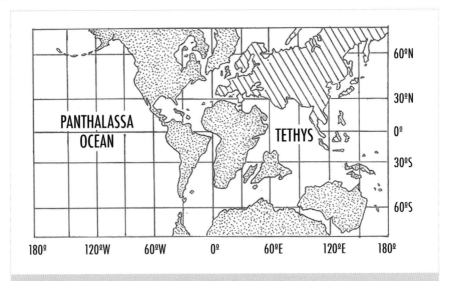

Figure 2
Earth's land masses about 100 million years ago. North and South America are separating from Africa to start to form the Atlantic. The future Antarctica is moving south, while Australia and the Indian subcontinent are moving northwards across the *Tethys* Sea.

But before this happened, by about 50 million years ago (see Fig 3) Africa was moving towards the western end of *Eurasia*, closing off the remains of *Tethys* to become the Mediterranean. At this time the Atlantic and Indian Oceans were pretty similar to the way they are now; and while the old *Panthalassa* was still huge, it was by now fairly similar to the present Pacific. The proto-Mediterranean, however, was in for a dramatic series of changes.

About 20 million years ago the eastern end of the Mediterranean became closed off from the Indian Ocean, a connection not to be made again until the Suez Canal was opened in 1869. Then, a little later, the western end was closed, separating it from the Atlantic. Over the next 14 million years or so, with no major rivers running into it and with high evaporation from its surface, the proto-Mediterranean gradually dried up. By 6 million years ago it was a huge, and largely dry, depression between southern Europe and North Africa, averaging 1,500 metres deep and with a few relatively small bodies of water in its bottom. Then, about 5.3 million years ago, one of the most dramatic events in the recent history of the earth happened; the Atlantic broke through into the almost dry Mediterranean, opening up the modern Strait of Gibraltar. There were, of course, no human beings about at the time, but what an awesome sight it would have been to the animals whose habitat was about to be obliterated. The resulting cataract was perhaps 3,000 metres high,

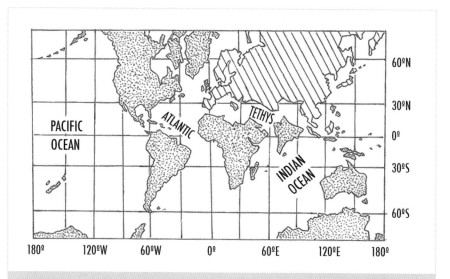

Figure 3
Land and sea distribution about 50 million years ago. India is now approaching Asia and the proto-Indian Ocean is more or less separated from the proto-Pacific by Australasia and Indonesia. As North and South America move away from Europe and Africa to widen the proto-Atlantic, the proto-Pacific, basically the remains of the vast Panthalassa, is becoming correspondingly narrower. This same process, with the Atlantic getting wider and the Pacific getting narrower, is still going on today – but very slowly.

50 times as high as Niagara, and poured almost 200 km³ of water each day into the vast hole, taking 50-60 years to fill it to the top as we see it today.

So the modern Mediterranean is by far the youngest of our present day oceans, but, paradoxically, it will also be the first to disappear. Africa is still moving north, colliding slowly but inexorably with Europe and pushing up the Alps in the process. Eventually, the Mediterranean will be completely obliterated while all the other modern oceans are still in existence. But the 300 million humans who live permanently around the Mediterranean, and the 110 million or so tourists who visit the area each year, don't need to worry too much because the Med will be with us for a few millions of years yet.

4 How salty are the oceans?

Pretty salty. On average about 35 grams, that is about two heaped tablespoons, in every litre. Around the mouths of rivers, the saltiness (or salinity as it is referred to by oceanographers) may be reduced by the inflowing fresh water. Indeed, because fresh water is much lighter than salt water, the surface layer around the mouths of big rivers may be almost fresh, overlying salt water underneath. Similarly, fairly enclosed seas may have reduced salinity because they have rivers flowing into them but not very much exchange of water with the open sea. The salinity of the Baltic, for example, ranges from almost 35 grams per litre near its opening to the North Sea to almost zero in the Gulf of Bothnia, while the surface waters of the Black Sea, fed by the major rivers Danube, Dneiper and Don, are about half as salty as full seawater.

At the other extreme, in areas of intense evaporation, such as the Red Sea and the Arabian Gulf, the salinity of the surface layers may exceed 40 grams per litre.

But the salinity of the vast majority of the water in the open ocean ranges within rather narrow limits, from about 33 grams per litre to a bit more than 37 grams per litre. Small though these variations are, they are extremely important because of their effects on the density (i.e. heaviness) of the water and therefore on ocean circulation.

The consistency of salinity in the world ocean is remarkable enough, but even more amazing is the fact that although seawater is a complex solution, probably containing all the naturally occurring elements on earth[2], the relative proportions of the major constituents, making up more than 99.9% of the total salinity, remain virtually identical throughout the oceans.

More than 90% of the salt in the sea, about 26,000 million million tonnes, is exactly the same as the stuff we put on our food, sodium chloride. But the oceans are so vast that even the 'rarer' elements are present in impressive quantities. For example, there are about 5.3 million tonnes of gold, 2.6 million tonnes of silver and even 92 tonnes of radium in the sea, but very thinly spread. So far attempts to extract them have proved to be more expensive than the stuff extracted!

5 Why is the sea salty?

Q4 explained how much salt there is in the sea and roughly what it contains. But exactly why is the sea so salty? A simple, and partly true, answer is that rain falling on the land flows into the sea through rivers and subterranean passages. As it does so it picks up small, but significant, amounts of salts and minerals from the underlying rocks and transports them to the sea. Water evaporates from the sea surface to form clouds, but leaves the salts behind. The clouds produce rain

---

2   Sadly including man-made combinations of them like DDT and PCBs.

and the cycle begins all over again. In addition, smaller amounts of minerals and salts are deposited into the oceans from volcanoes and from the land as wind-borne particles.

On the face of it this seems a plausible explanation, but it has two huge flaws. Firstly, if this were the complete story the oceans should be getting saltier and saltier. In fact, all the evidence suggests that the salinity has stayed more or less the same for many tens of millions of years, at least. How come?

Secondly, the relative proportions of the various elements in seawater are totally different from those in the fresh water from rivers that supply them. For example, sodium and chlorine, the components of common salt, represent almost 90% of the salts in the sea, but only about 16% of the salts in most river water. Similarly, rivers carry more calcium than chlorine, but the oceans contain about 46 times as much chlorine (in the form of chloride) as calcium. And while silica is a significant constituent of river water it is relatively much less abundant in the sea. Clearly, the situation is much more complicated than our simple explanation would suggest.

The true situation is extremely complex and many aspects are still not fully understood. But in principle the chemistry of the oceans is close to what chemists call a *steady state*. This means that the inputs of chemicals from rivers, winds, volcanoes and so on are almost exactly balanced by the processes removing them from the oceans. Some of these processes are purely chemical or physical, such as the sinking of volcanic ash or sand particles to the ocean floor, or the accumulation of elements into metalliferous (that is metal rich) muds or manganese nodules. In addition, many of the ocean chemicals are involved in biological processes. Some biologically important chemicals are used in very small quantities or are recycled and used over and over again. But some make a more or less one way journey into the biological systems and then completely out of the oceans for ever, or at least for a very long time. For example, calcium and silicon are both used by many tiny marine organisms as major constituents of their skeletons. When these organisms die, as gazillions do each year, their skeletons (and the contained calcium and silicon) sink to the deep sea floor where they form the kilometres thick carpet of sediment and may be locked up for thousands or millions of years.

Finally, some constituents, notably chlorine, play virtually no part in the geological or biological processes. Since it is supplied in only minute quantities in the incoming river water, it appears that the chlorine in the present day oceans must have been there for a very long time, obtained from the earth's primitive atmosphere when the water vapour first condensed to form liquid seas.

6 How deep are the deepest parts of the oceans?

The short answer is 10,924 metres, which is equal to 35,842 feet or almost 7 miles. This depth, the greatest ever recorded and probably the most accurate, was measured in 1984 by a multibeam echo sounder on the Japanese survey ship *Takuyo*, in the Challenger Deep in the Marianas Trench, which is in the south-western Pacific. It is the most recent of a whole series of 'record' depth measurements made in this area of the oceans since the Royal Naval ship HMS *Challenger* first found it in 1875 and obtained a sounding, that is a depth measurement, of 4,475 fathoms (26,850 feet or 8,150 metres) using the then normal method of lowering a heavy weight to the sea floor on the end of a long rope.

It is possible, but rather unlikely, that slightly deeper spots will be found in the future. But two things are certain; first, that any very deep spot will be found within a deep sea trench (see Q16), that is one of a series of narrow, steep-sided gashes in the sea floor almost certainly, like the Marianas Trench, in the western Pacific, and second, that any increase in the record will be by no more than a few metres. Nevertheless, even as the record stands now, it is interesting to think that Mount Everest could theoretically be dumped in the Marianas Trench and have almost one mile of water over its summit!

On the other hand, it is equally intriguing to think that the bottom of the Marianas Trench and the summit of Mount Everest are separated vertically by only about 65,000 feet, that is just over 12 miles or 20 km. Since the earth has a mean diameter of some 7,917 miles or 12,742 km, this extreme 'roughness' of the planet's surface represents only 0.15% or about 1/640th of the diameter. Standard billiard balls are 52.5 mm in diameter and must be made to a tolerance of plus or minus 0.05 mm. In other words, an individual ball can have diameters ranging between 52.45 mm and 55.55 mm. The difference between these two extremes, one tenth of a millimetre equals 1/525th of the mean diameter; that is, relatively slightly bigger than the roughness of the earth. This is not quite the same as saying that the earth is as smooth as a billiard ball, since a billiard ball with a pimple on its surface the relative size of Mount Everest would certainly not pass muster. Nevertheless, it is interesting to think that the ups and downs of the earth's surface in an exact replica made the same size as a billiard ball would fall within the tolerance ranges of a billiard ball's diameter!

Of course, the replica wouldn't pass the billiard ball inspector's scrutiny for one other reason, the fact that the earth is actually an oblate sphere, slightly flattened from pole to pole (see Q80). A billiard ball of this shape would behave very strangely indeed!

7 How does pressure change with depth in the sea?

Every land animal on earth lives under a pressure caused by the weight of the overlying atmosphere. At sea level, although this atmospheric pressure varies a bit from place to place and from time to time, it is generally about 15 pounds per square inch or one kilogram per square cm. This is known as one atmosphere and, of course, it operates in all directions, that is sideways and upwards as well as downwards.

A similar effect occurs in water, but because water is much heavier than air, the pressure increases much more rapidly with increasing depth in the sea. The precise weight of seawater, and therefore the rate at which the pressure increases with depth, depends on the water's temperature and its salt content or salinity (see Q4) so, like atmospheric pressure, it varies a bit from place to place and from time to time. But for most purposes it is close enough to assume that a column of seawater about 10 metres or 33 feet deep weighs just about the same as a column of air of the same cross-sectional area extending from the earth's surface to the very limits of the atmosphere.

Consequently, pressure in the sea increases by one atmosphere for approximately every 10 metres of increasing depth. So at 10 metres down the pressure is twice that at the surface, at 20 metres three times, at 100 metres eleven times and so on. This means that at 1,000 metres down the pressure is about 100 atmospheres, that is 1,500 pounds per square inch or 100 kilos per square cm, while at the bottom of the Marianas Trench (see Q6) at a depth of almost 11 km the pressure is a staggering 16,500 pounds, nearly 7.5 tons, per square inch, or more than a metric tonne on each square cm! That's about like balancing a large car on the tip of one of your fingers.

And yet all sorts of animals live in the sea all the way down to these depths (see Q29). So how on earth do they do it without being squashed to a jelly? Have a look at Q30 to find out why.

8 Why is the sea blue?

The short answer is that it is simply because water in the mass is, indeed, blue rather than the clear colourless stuff it appears to be in a glass. But this needs a bit of explanation.

If you pose the blue sea question to a pretty average group of adults, as I have several times, the most common answer you hear is that the sea reflects the colour of the sky. But this can't possibly be true. If it were, the sea should appear white when the sky is obscured by white fluffy clouds, which it clearly doesn't, while it is often anything but blue even under a very clear blue sky. In high latitudes, in both the northern and southern hemisphere, the sea may look reasonably blue, though often rather dark and slaty in deep open ocean areas. In shallow inshore waters, on the other hand, it is usually green or brownish, sometimes a muddy yellow or even red. These colours are often assumed to be due to impurities in the water making it in some way 'dirty'. This may be true, but the water colour is more often due to the presence of *phytoplankton*, tiny plants called algae that contain chlorophyll, like land plants.

Chlorophyll absorbs blue and red light and reflects yellow-green light, giving the sea that mucky sort of appearance but often indicating that it is very productive (see Q28). Some of the tiny single celled algae, the coccolithophores, have chalky calcareous shells and can occur in such vast numbers that they give the sea surface a creamy white appearance that is visible from space. But water without such particles in it can, indeed, look quite blue. And the fewer the particles, the more blue it looks.

The reason behind all of this is that light from the sun arriving at the earth includes light of lots of different wavelengths, or visible colours, from short wavelength blue light to long wavelength red light[3]. In fact, what we call 'white light' contains all the colours of the rainbow, so when we see coloured objects, say a red book or a yellow flower, it is because they reflect to our eyes these particular parts of the spectrum and absorb the others. In exactly the same way, when sunlight passes through pure water with no suspended particles, the water molecules absorb the red end of the spectrum and scatter the blue light in all directions. Consequently, when some of this scattered light reaches our eyes the water looks blue. And since, as Q28 explains, it is the waters in the warm central ocean regions that contain the fewest particles, these are the bluest!

9 Why do icebergs float?

Icebergs are lumps of ice broken off the ends of glaciers and ice shelves as they reach the sea. Glaciers, whether in the Arctic or the Antarctic, are made up of snow that has fallen on high ground tens or hundreds of thousands of years previously and has been converted into ice by compaction under the huge pressure of

---

3   Sunlight also contains even shorter wavelength ultraviolet light and longer wavelength infrared light, but since our eyes can't detect these wavelengths they are not relevant to this problem.

subsequent eons of snowfall accumulated above it. Despite the apparent solidity of ice, a bit like glass or even metal, the enormous weight of new ice above the older stuff causes it to become semi-liquid so that the glaciers flow slowly down gradient towards the sea.

Since the original snow, like rain, fell as fresh water, the ice produced from it is also fresh and is therefore considerably lighter than polar seawater which has 33–34 grams of salt in every litre (see Q4). So the glacier ice floats!

But there are two other reasons why it floats. The first is that the snow from which the glacier ice was formed in the first place contained a considerable amount of air. Some of the entrapped air would have been lost as the snow started to become compacted, but much of it remains in the ice as tiny air bubbles helping to make the ice lighter than it would otherwise be, although not by much. The second and more significant reason is that fresh water and salt water behave quite differently when they are chilled. Most materials shrink with decreasing temperature and therefore get heavier. Water is no exception, but it has some very strange peculiarities. If you chill seawater, it behaves exactly as you might expect, becoming heavier and heavier as the temperature falls right up to the point when it freezes, though this doesn't happen until well below zero degrees centigrade.

In contrast, if you chill fresh water it contracts and becomes heavier and heavier until the temperature reaches just under 4°C. But if you cool it further, it actually starts to expand and becomes lighter and lighter until it freezes at very close to 0°C. This is why garden ponds freeze at the surface first, and fish are reasonably happy swimming at the bottom in relatively warm 4°C water. Only if the air temperature stays very low for a long period will the pond freeze solid, partly because the layer of ice at the surface acts as an insulator between the cold air above and the warmer water below.

So freshwater ice will float in unfrozen fresh water and even higher in salt water. In fact, icebergs have about one ninth of their mass above water and eight ninths submerged.

10 Why are icebergs white?

Q8 explained that pure water with no suspended particles in it looks blue because the water molecules absorb red light and only scatter (i.e. reflect) blue light. So since glaciers are simply frozen water, albeit fresh water, shouldn't they look blue too?

Well yes they would, if they were simply frozen water. In fact, pure frozen seawater sometimes becomes added to the bottom of deep icebergs and this 'marine ice'[4] is, indeed, a deep blue colour. But it is usually only visible if the iceberg tips over as it melts to bring this originally deep layer above the surface.

But the bulk of an iceberg is not produced in this way. Q9 pointed out that icebergs, in both the Arctic and Antarctic, are made of ice formed from snow that fell many years previously. When it did so, lots of air was trapped between the snowflakes and, as it became compacted into ice, some of the air remained locked in it as tiny air bubbles. It is these air bubbles that are responsible for the white appearance of the icebergs because they act as tiny mirrors reflecting (or rather scattering) all the colours of the spectrum which, together, make up white light. In exactly the same way, tiny water droplets in the air, forming clouds, also scatter all visible wavelengths – which is why clouds look white.

11 Why are icebergs in the Arctic and Antarctic so different?

The answer in a word, or rather two words, is ice shelves.

Arctic icebergs are bits broken off the ends of glaciers where they meet the sea in a process called calving. The resulting lumps of ice tend to be rather irregularly shaped and, although they may be tens of metres across and weigh many hundreds or thousands of tonnes, they are much smaller than many of their Antarctic cousins. Incidentally, in order to qualify technically as icebergs they need to stick out of the water by at least 5 metres. At the next size level down, bits of floating ice are rather endearingly called 'bergy bits', while if they are no more than a couple of metres across they will be called 'growlers' because of the noise they make when they grind together.

But no matter how big or small the Arctic icebergs are, like all freshwater ice (see Q9) they will float with about one ninth of their mass above the water and the other eight ninths below. As the underwater parts melt (because the seawater is often warmer than the overlying air), the floating lumps of ice become top-heavy and eventually tip over until a new balance is found with, as before, one ninth above and eight ninths below the waterline. In this process, icebergs

---

4   This ice is called 'marine ice' to distinguish it from 'sea ice' which is frozen seawater formed at the surface. Sea ice contains at least some air bubbles and tends to look somewhat milky rather than white.

often become sculptured into fantastic shapes, as if made by some supernatural artist.

The same process happens in the Antarctic, of course. But whereas the Arctic icebergs, and some Antarctic ones, come from glaciers that are relatively narrow, flowing down to the sea via rather restricted valleys, the main producers of Antarctic icebergs are the continent's huge ice shelves, the seaward extensions of the Antarctic ice sheet.

The ice sheet is the largest mass of ice on earth and overlies almost the whole of the Antarctic continent. It is the result of hundreds of thousands of years' snowfall which, under the weight of successive layers, has become compressed into a sheet of solid ice with an average thickness of 2,700 metres. The huge weight of ice actually depresses the underlying land, but it also causes the ice itself to flow outwards all around the continent, a bit like icing poured onto the middle of a Christmas cake will flow towards the cake's rim. Instead of running down the edge of the cake, the Antarctic ice sheet spreads over the surface of the adjacent sea to form ice shelves, particularly in three major areas: the Filchner Ice Shelf in the Weddell Sea to the south of the Atlantic, the Ross Ice Shelf in the Ross Sea south of the Pacific, and the much smaller Amery Ice Shelf south of the Indian Ocean.

The ice shelves are much thinner than the main ice sheet, but can be several hundreds of metres thick, and their huge weight is supported by the seawater as they extend out to sea. As the lower parts melt, the overlying weight causes great sections to break off to produce the massive flat-topped tabular icebergs so characteristic of the Southern Ocean. These are typically 30 or 40 metres high above the water but extend as much as 300 metres below the waterline. They are often tens of kilometres long and may even reach 100 km or more. They sometimes run aground in shallow areas and become semi-permanent features of the seascape, effectively becoming islands staying in place for as much as ten or twenty years.

Eventually, even these goliaths will melt, particularly if they are carried by water currents northwards towards warmer waters. As they do so, just as with smaller icebergs, they will tip over as the weight distribution changes and lose their tabular shape. At this stage they are especially dangerous for mariners because they may have huge unseen hull-ripping spikes extending tens of metres underwater beyond the boundaries of the visible above-water parts. On a much smaller scale, it was one of these underwater ice scythes on a north Atlantic iceberg that tore open the *Titanic* with such tragic consequences on that fateful April night in 1912.

12 What is sea ice?

Sea ice, as its name suggests, is ice formed when seawater freezes. In contrast to the origin of icebergs from glaciers or ice shelves (see Q9), the water from which sea ice is formed is already salty. As the ice forms, the salts are excluded from it in a process called brine rejection so that the resulting ice is relatively fresh and, like true freshwater ice, is therefore lighter than seawater. Hence it floats; but because it is not quite as salt free as iceberg ice, nor does it include so much air, sea ice floats a little deeper in the water with a slightly smaller proportion projecting above the surface.

13 What are the Continental Shelves?

Continental Shelves are the shallow parts of the sea floor adjacent to the main land masses or continents, hence the name. Geologically, they are seaward extensions of dry land and their topography, or ups and downs, tends to reflect that of the adjacent land. But the average seaward slope tends to be rather gentle, generally no more than about 1° or 1 in 60. In contrast, beyond the seaward limit of the Continental Shelf, known as the Shelf Break, the slope of the sea floor typically increases to about 4° or 1 in 15 or so (see Q14).

The width of the Continental Shelf can vary from just a few kilometres, for instance off the coast of Portugal, to more than 1,500 km. For example, the whole of the British Isles stands on the north-western European Shelf which therefore forms the seabed beneath the North Sea, the Irish Sea and the English Channel.

The outer edge of the Continental Shelf, the Shelf Break, lies at a depth of about 200 metres in the north-eastern Atlantic, but worldwide its depth ranges from as little as 20 metres to as much as 500 metres, with an average at about 130 metres.

The Continental Shelves together make up only about 6% or one twentieth of the total surface area of the world ocean (see Fig 4), yet they account for something like 99% of the total living resources, including fish and shellfish, that we get from the seas. This is partly because the Continental Shelves are the nearest parts of the oceans to the world's fishing ports. But it is also because they support the richest growth of microscopic marine plants, the phytoplankton (Q27), and in turn the zooplankton (Q26) and fish.

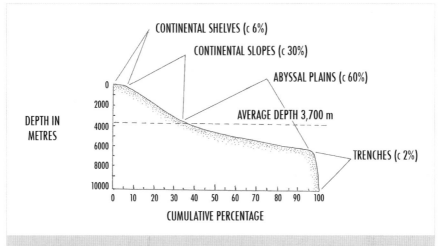

Figure 4
Percentage distribution of different depth zones in the ocean. Note that by far the largest part of the oceans cover the Abyssal Plains at depths between about 4,000 and 6,000 metres. The Abyssal Plains make up about 60% of the surface of the oceans and more than 40% of the total surface of the earth!

14 What are the Continental Slopes?

As the name suggests, Continental Slopes are those parts of the ocean floor linking the relatively shallow Continental Shelves (Q13) and the Abyssal Plains (Q15). At the shallow end, Continental Slopes begin at the Shelf Break (Q13), mostly at a depth of about 200 metres (c660 feet), and end at a depth of 3,000 to 4,000 metres (9,800 to 13,100 feet) as the seabed flattens out onto the Abyssal Plain. The slope of the bottom over this region is usually about 4° or 1 in 15, the equivalent of a fairly steep hill for a cyclist. So to get from a depth of about 200 metres to 3,000–4,000 metres the continental slopes are generally about 40 to 60 km wide and together represent between 20 and 30% of the total surface area of the world ocean (see Fig 4)[5]. But in some areas the continental slope is much narrower and, as a result, steeper. Indeed, the surface of the underlying rock may be almost vertical. Finally, in some areas, including to the south west of Britain, the continental slope is quite wide, but is bisected by steep-sided canyons so that although the general slope is not particularly steep there are some extremely precipitous areas within it.

5   Note that the 'Continental Slope' part of this diagram includes the flanks of the mid-ocean ridges (see Q17) that are in the same depth range as the true Continental Slopes but are totally separated from them.

15 What are the Abyssal Plains?

Abyssal Plains are the largest and flattest places on earth. They make up the seafloor beneath almost two thirds of the surface area of the world ocean, and nearly half of the total surface area of the earth (see Fig 4). Consequently, they represent by far the largest biological environment on the planet.

The Abyssal Plains extend between the foot of the Continental Slopes at a depth of 3,000 to 4,000 metres and either the foothills of the mid ocean ridges (see Q17) or the edges of the deep ocean trenches (Q16) at a depth of about 6,000 metres. The slope of the Abyssal Plains is typically less than 1 in 1,000, though they may be interrupted by abyssal hills, often rising some hundreds of metres from the deep sea floor.

16 What are Deep Sea Trenches?

The deep sea trenches are special areas, called subduction zones, where one of the earth's tectonic plates (Q22) is sliding beneath another and dragging the sea floor down with it into a long and narrow sort of wrinkle. The deepest trenches are concentrated in the western Pacific from north of Japan down to New Zealand, with rather shallower ones in the western central Atlantic.

The trenches are characterized by water depths greater than about 6,000 metres (19,700 feet) and collectively represent less than one fiftieth (less than 2%) of the total surface area of the oceans.

At the other end of the depth scale, the Continental Shelves, that is the relatively shallow parts of the oceans bordering the great land masses (see Q14), have depths generally less than about 200 metres (650 feet) and cover only about one twentieth (5-6%) of the world ocean (see Fig 4). So the vast majority of the oceans cover intermediate depths including the vast Abyssal Plains (Q15) at depths between 3,000 and 6,000 metres (9,800 and 19,700 feet). Put all this information together and you come up with an average depth for the world ocean of 3,700 metres, that is 12,150 feet or almost two and a half miles. This seems, and is, a huge depth of water. But to put it into the context of the earth as a whole, it represents less than one three-thousandth part of the diameter of the planet. So our world, from the tops of the highest mountains and down to the bottom of the oceans, represents an extremely thin skin on the surface of the earth. This is why, on photos of the whole earth taken from space, we notice the land masses and the oceans but not the mountains. At this scale, the earth looks just about as smooth as a billiard ball – and indeed it is (see Q6).

17 Is the bottom of the deep sea flat?

A lot of it is, but quite a bit isn't. To illustrate this, let's take an imaginary walk across a fairly typical piece of the deep ocean and see what sort of terrain we encounter on the way. Although the fine details would vary depending on our particular route, the general features would be very much the same whether we are crossing the Indian, Pacific or Atlantic Oceans; but since the North Atlantic is the one that I am most familiar with, let's make our imaginary trip from the south-western tip of the United Kingdom at Land's End across the Atlantic towards Newfoundland, a distance of some 1,800 nautical miles (Q80) or 3,400 km (see Fig 5). Naturally, we need to go only half way across because by then we will have seen more or less all that the ocean floor has to offer and the rest of the journey would be more or less a mirror image of the first part. On the other hand, we mustn't stop only a quarter or a third of the way across because, if we did, we would miss one of earth's most amazing features, a mountain range that knocks all the others into the proverbial cocked hat. So let's go.

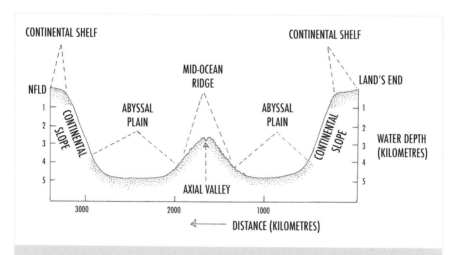

Figure 5
Cross section of the sea floor of the North Atlantic from Land's End on the right to Newfoundland on the left. Note that the vertical scale (a maximum of about 5 km for the depth of the sea) is 200 times bigger than the horizontal scale (almost 3,500 km indicating distance across the ocean). So this illustration hugely exaggerates the topography of the seabed and makes the slopes look much steeper than they really are. If the vertical scale on this diagram was the same as the horizontal one, all of the sea floor irregularities would be 'lost' within the horizontal line marking the distance scale (see Q6). On the other hand, if the diagram had been drawn with both scales the same as the vertical one, the words NFLD and Land's End would have been about 15 metres apart!

As we walk from the tide-line into deeper and deeper water, our last view of the land would probably be a seaweed-covered rocky shore, so typical of the western British Isles. We might well encounter odd patches of rock as we move offshore, but generally the bottom will be dominated by sediments ranging from fairly coarse gravels, through sands to fine muds. The finer sediments will be mainly confined to deeper areas or depressions in the sea floor, less affected by currents than the other regions. As we move across the inshore shallow area – the Continental Shelf (Q13) – the bottom may undulate quite a bit, but the average offshore slope will be fairly gentle until we reach the Shelf Break at a depth of 200 metres or thereabouts. The break is by no means a clearly defined line, but as we move across it we would be aware that the downhill slope is gradually becoming steeper, from something like 1 in 100 to 1 in 15 or so.

We are now moving down the Continental Slope (Q14), and while we might come across the odd rock outcrop, just as we did on the Continental Shelf, the bottom sediment will generally be even finer than before, now dominated by creamish-coloured muds and silts. Eventually, at a depth of around 3,000 metres, the slope will decrease to something like 1 in 100, that is about the same as on the Continental Shelf. This is the Continental Rise, a gentle sort of ramp linking the Continental Slope with the largest environment on earth, the Abyssal Plains (Q15).

Between them, the world's Continental Shelves, Continental Slopes and Continental Rises together underlie about one third of the total surface area of the oceans, a bit more than 100 million square km. At the other end of the depth spectrum, the deep sea Trenches (Q16), at depths between 6,000 metres and 11,000 metres, occupy less than one fiftieth or 2% of the total ocean surface area[6]. In the middle come the Abyssal Plains, covering most, but not quite all, of the rest.

Because of the eternal darkness at the bottom of the ocean, nobody has ever seen an Abyssal Plain, other than the tiny bit of it visible through the port hole of a submersible or on deep sea photographs. But if you could see it, it would be a truly awesome sight. Covered in a blanket of fine mud, kilometres thick, the sea floor would stretch away as far as the eye could see, apparently as flat as a billiard table. Apart from local ups and downs of a few centimetres caused by animals crawling in and over the mud, the Abyssal Plains are the flattest places on earth, with slopes of 1 in 1,000 or more sometimes stretching uninterrupted for hundreds if not thousands of km, an almost unimaginably monotonous sight. Nevertheless, even the plains are not

6   We won't come across a Trench during our journey because they are confined to the western parts of the oceans, around the Caribbean in the Atlantic and, in the Pacific, stretching along a great arc from New Zealand in the south to Japan in the north. But if we did walk across one we would be struck by how long and narrow they are, just a few km across, but tens of km long, and with steep sides plunging down to the trench floor at horrific slopes of 1 in 5 or sometimes even steeper.

without interruption, for scattered throughout the world's oceans are submerged hills rising abruptly through the bottom muds like huge carbuncles. By convention, if these hills rise more than 1,000 metres above the abyssal sea floor they are called seamounts, while if they are less than 1,000 metres high they are abyssal hills. Either way, they are mostly the remains of old volcanoes and are more abundant in the much more seismically active Pacific than in the Atlantic and Indian oceans.

But even in the Pacific, these bumps in the abyssal plain are fairly few and far between, except near the vast mountain chain making up the mid-ocean ridge system. As the name suggests, in the Atlantic and Indian Oceans the ridge runs more or less down the middle, though in the Pacific it is skewed towards the east (see Fig 9). Linked together, the ridges form a 45,000 km (28,000 mile) long continuous mountain range rising 3,000 metres or more above the Abyssal Plains. They mark the boundaries between the earth's tectonic plates (Q22) and are the result of the extrusion of molten magma from many km beneath the seabed to form new sea floor which is then forced away from the ridge on either side. This sea floor spreading is at a rate of about 1 to 2 cm a year in the Atlantic (about the same rate that fingernails grow) and several times faster in the Pacific[7].

Because the sea floor produced at the mid-ocean ridges is new, as we approach the ridge we are moving across younger and younger material. Consequently, the sediment blanket covering it becomes thinner and thinner because it has had less time to accumulate. Eventually, as we approach the foothills of the ridge system, outcrops of rock begin to protrude through the sediment cover, and as we climb higher and higher up the slopes of the ridge, the sediment cover becomes less and less prevalent and the exposed rock correspondingly more common. Finally, at the very peak of the ridge in the Atlantic, we will find a pair of parallel ridges separated by an axial valley, typically a depression 100–200 metres deep and up to 10 km wide. On the slower spreading ridges in the Pacific there is often no axial valley. But valley or not, the mid-ocean ridges are the most volcanically active continuous zone on earth and it is here, as we cross the top of the ridge, that we may be fortunate enough to come across one of the amazing communities surrounding a hydrothermal vent (see Q18).

The ridge crest, and most hydrothermal vents, are generally well beneath the surface of the sea, at depths between about 2,000 and 3,000 metres, that is about level with the middle to lower parts of the Continental Slopes. Occasionally, however,

---

7   Despite the faster production of new sea floor in the Pacific than in the Atlantic, the Pacific is currently getting smaller and the Atlantic is still getting bigger. This is because the Pacific sea floor is being lost (that is 'subducted' beneath the American and Asian plates) faster than the Atlantic is. Since the earth as a whole seems to be neither increasing nor decreasing in overall size, the processes of sea floor production and loss must be more or less balanced.

small parts of the ridge extend upwards as huge submarine mountains with their peaks emerging through the ocean surface. Examples in the Atlantic are Tristan da Cunha and Ascension Island in the southern hemisphere and the Azores in the north. Even less commonly, much larger sections of the ridge system emerge from the ocean depths and give us a more direct glimpse into the inner workings of the earth's renewal system. The most dramatic of these is Iceland, which is actually bisected by the North Atlantic mid-ocean ridge, that particular bit called the Reykjanes Ridge. The Reykjanes Ridge separates the American Plate and the Eurasian Plate, so Iceland actually sits with one foot in each camp and thanks to this split personality, Icelanders benefit from free geothermal energy from their own personal hydrothermal vents. Of course, such a situation comes with its downside in the form of constant risks of fairly violent volcanism.

In recent years, the most graphic example was the emergence of what became known as Surtsey Island (named after the Norse fire god Surtr). Surtsey appeared between November 1963 and June 1964 as the result of a volcanic eruption on the Reykjanes Ridge from some 130 metres beneath the surface of the sea to the south west of Iceland. By the end of this period the new island had attained a surface area of about 2.7 square km, though erosion by wind and waves since then has reduced it to about half that area. Surtsey was declared a nature reserve in 1965 while the volcano was still active and, since that time, has provided scientists with a priceless natural experiment into the way in which plants and animals colonise new areas.

So that is just about the end of our journey. Having reached the top of the mid-ocean ridge system we have seen pretty well everything that the sea floor has to offer. Clearly, the answer to the original question about whether the bottom of the deep sea is flat is 'Yes, in parts', a bit like the curate's egg!

18 What are Hydrothermal Vents?

Hydrothermal vents are holes or cracks in the deep sea floor through which flow plumes of superheated water carrying a complex cocktail of chemicals picked up from the underlying sediments.

The vents are concentrated near the edges of tectonic plates (see Q22) and particularly on the mid-oceanic ridges. These are areas where new sea floor is produced as the plates are separating, allowing molten rock or magma to rise from the earth's mantle through the relatively thin overlying crust to solidify into glassy basalt rock as it cools. This new sea floor is pushed away from the ridge axis on either side as new molten material arrives at the surface.

In the cooling process, lots of fissures and fractures are produced in the solidifying rocks and these provide conduits for the hot water. The water itself is normal seawater that has seeped through the seabed sediments (see Fig 6), sometimes over considerable distances, and has penetrated deep beneath the sediment-water interface, sometimes several km. Here it comes into contact with hot basalt produced from the magma and is heated to temperatures of 350–400°C. If this happened on land the result would be catastrophic as the water boiled and turned to steam as the temperature reached about 100°C. But most vents are situated at depths between about 2,000 and 4,000 metres where the water pressure is 200–400 times higher than atmospheric pressure (see Q7) so, despite the enormous temperatures, the water doesn't boil.

However, it does become very light and rises rapidly towards the surface of the sea floor. As it does so, it loses some of its dissolved constituents (particularly magnesium, oxygen and sulphate ions) and these are replaced by others leached from the rocks, including iron, manganese, copper, zinc and lead – and lots of sulphide ions. As the hot water approaches the surface of the sea floor and mixes with the ambient cold (about 2°C), well oxygenated deep sea water, the rising plume water's load of chemicals is instantaneously precipitated as a cloud of metal rich sulphide and oxide grains giving the water the appearance of smoke. At very high temperatures of around 350°C or more the particles in suspension are mostly black sulphides producing 'black smokers', while at slightly lower temperatures (below about 330°C) they are largely lighter coloured sulphates and oxides, resulting in 'white smokers'. In both cases the minerals are deposited around the emerging plumes to form tubular chimneys that may grow into impressive pillars several metres high before they eventually become unstable and topple over.

Despite the high temperature of the plume water and the speed with which it emerges from the vents (up to 5 metres a second), they are surrounded by so much cold water that the temperature falls to not much above ambient (i.e. about 2–3°C) within a few metres of the vent opening. Nevertheless, the plumes may be detectable several hundreds of metres above the seabed using modern instruments capable of detecting temperature differences of a few hundredths of a degree.

Apart from this amazing gushing of superheated water from the sea floor, hydrothermal vents are also the home of many remarkable animal communities found nowhere else on earth. They were first discovered in the eastern Pacific in the 1970s and some 200 active vents have subsequently been found at many sites on the oceanic ridge system in the Pacific, Atlantic and Indian Oceans.

The animals living around the vents are extremely unusual because, whereas virtually all the other animals living in the oceans are dependent on plant production in the sunlit near-surface zone as a result of photosynthesis (see Q24 and Q27), these vent communities are totally independent of sunlight. Instead, they

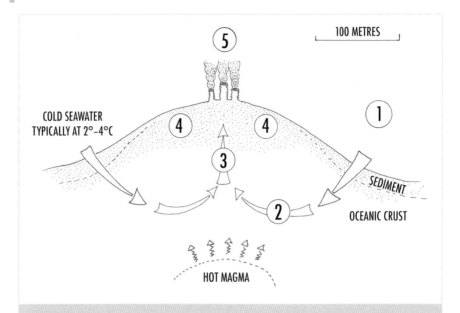

Figure 6
Cross section through a fairly typical hydrothermal vent field.

1: Cold seawater seeps down into the seabed through the sediment and then through cracks in the ocean crust.
2: The water is heated by the basalt produced by the underlying magma and picks up a cocktail of chemicals.
3: The hot and buoyant water rises back towards the seabed surface.
4: As it cools, the rising water loses some of its chemical load, particularly metal sulphides, which are deposited in the sea floor.
5: Water gushes through the sea floor at temperatures up to 350°C or more, but mixes with the surrounding cold water and rapidly cools. As it does so, it deposits sulphide 'chimneys' that can reach heights of tens of metres.

are based on a quite different process called chemosynthesis, which does not use sunlight as its energy source.

The key to the chemosynthetic systems is a number of specialised deep-sea bacteria that are able to extract energy from the cocktail of chemicals, particularly sulphide containing ones, in the vent water as it emerges from the sea floor. Using this energy to make complex molecules, much as green plants use energy from sunlight, the bacteria form huge mats of material fed upon by almost equally specialised vent animals. Because the chemosynthetic bacteria are able to grow so rapidly, the vent communities can be extremely rich, sometimes hundreds or thousands of times more abundant than the populations of non-vent animals living at comparable depths.

Since the first vents were discovered, more than 500 vent species, previously totally unknown, have been described – that is an average of one new species every couple of weeks! All of them are remarkable, but some are totally gobsmacking. When the first Pacific vents were investigated, biologists around the world were amazed to learn of the existence of colonies of worms, each up to two metres long and 5 cm in diameter; if this wasn't incredible enough, they turned out to have neither mouth nor anus, but cultivated the bacteria actually inside their bodies. Almost as remarkable were giant clams 30 cm or more long, also living symbiotically[8] with vent bacteria, but in this case on the surfaces of their gills.

Different ridge systems seem to have quite different animal communities, and when the Atlantic ones were discovered, instead of giant worms and clams, scientists found huge swarms of shrimps seething around the vents in their tens of thousands like bees round a vast jam pot. Like quite a few deep sea creatures, these shrimps lacked conventional eyes. Instead, their eyes appeared to have developed a quite different sensitivity – to *infra-red* radiation. Such radiation is rare indeed in the deep sea – except that it is exactly what is produced by very hot water! So it seems that the vent shrimp can 'see' the gushing water even in the total blackness of the deep ocean. And when the behaviour of these animals was looked at in more detail it turned out that their 'swarming' around the vents is a strategy to help keep each individual within a fairly narrow temperature range around 35°C. There is a very good reason for this. The shrimps, like the big worms and the giant clams, 'farm' their own bacteria, in this case mostly on the inside of the carapace, that is the shell that covers the front half. Periodically, the shrimp reach inside their shells with their tiny pincers to crop the bacteria and transfer them to their mouths. Surprise, surprise, the bacteria grow best at exactly the temperature the shrimps keep swimming in!

These vent communities have been, and continue to be, the subject of enormous research effort. Not only are they fascinating for their own sake, but they have two other major interests. First, living as they do in one of earth's most spectacularly hostile environments, they may have lots of lessons for humans dealing with similarly difficult situations, including medical ones. Second, some scientists believe that hydrothermal vents have been around in one form or another since before life existed on earth. In fact, they suggest, life may have arisen first in these seemingly hostile conditions and only later developed a system based on sunlight. If so, the study of hydrothermal vent communities may represent, for biologists, an analogous situation to the way CERN enables physicists to investigate the origins of the universe, trying to determine what happened in the first few microseconds after the big bang from what they can see today.

---

8   Symbiosis simply means 'living together' and is a term used by biologists to describe a situation in which two different species co-exist in a relationship that is mutually advantageous.

19 What is Coriolis Force?

This is a difficult concept, but it is so important in understanding the oceans and the atmosphere that it is worth spending a few minutes grappling with it. So please bear with me.

To start confusing matters at the outset, Coriolis Force is not actually a real force at all. A proper 'force', in physics at least, is something that pushes or pulls on something else and either makes it move or makes it change its direction. The most familiar force, but one we generally take for granted, is, of course, gravity. When you drop a ball (or a priceless china pot!) and it falls to the ground it is because the force of gravity is pulling it towards the centre of the earth. When you push a lawnmower or pull a sledge you are providing the forces that cause them to move, while when the wind blows leaves off a tree, the wind is providing the force.

Coriolis Force is different. It is an 'apparent' force, resulting from the earth's rotation that 'appears' to deflect winds and water currents as they move over the surface of the earth so that they follow curved paths instead of straight lines. It also deflects anything moving across the earth's surface with little or no frictional contact with the earth, like a bullet, a missile or even a plane. Let's see how.

The earth rotates from west to east on its polar axis at the rate of one complete rotation every 24 hours. Since the earth's circumference at the equator is some 40,064 km or 21,633 nautical miles (nm)[9], this means that all points on the equator are moving consistently to the east at about 1,670 km or 901 nm an hour. People living at the equator (or anywhere else on earth) don't notice this movement, of course, because their whole world, ground, trees, buildings, atmosphere and so on, is moving along with them. And all of it is held firmly in place by dear old gravity!

As you move away from the equator, either to the north or the south, the circumference of a slice through the earth parallel to the equator decreases, slowly at first and then more rapidly. So at 10° north or south of the equator the 'slice circumference' has decreased, but only to about 39,463 km or 21,308 nm, and the easterly velocity has therefore dropped to 1,644 km or 888 nm an hour. By 45° north or south (the latitude of Bordeaux in the north and Dunedin (almost) in the south) the slice circumference has decreased to 30,877 km or 16,672 nm

---

9   Nautical miles rather than statute miles are used in these calculations since these are the inter-nationally agreed units for expressing distances and velocities over the earth's surface. Q80 explains the relationship between nautical miles, statute miles and kilometres.

and the easterly velocity is now 1,286 km or 695 nm an hour. But notice that although this is halfway between the equator and the North Pole, the slice circumference and the eastward velocity are still more than three quarters their values at the equator. But from now on these values decrease faster and faster as you move north (or south in the Southern Hemisphere). The halfway point in slice circumference and easterly velocity, that is about 835 km or 451 nm an hour, is reached at 60° north and south (the latitude of the Shetlands and Bergen in the north and well south of Cape Horn in the southern hemisphere), two thirds of the distance from the equator to the poles. Move 10° further north or south (to 70°) and the velocity has dropped to 571 km (308 nm) an hour, while at a latitude of 80° it is only 290 km (156 nm) an hour, about one sixth of the velocity at the equator. Finally, over the next 10° of latitude the rotational velocity falls to zero at the poles. Because of this varying situation, and somewhat counter-intuitively, the effect of Coriolis Force is at a minimum (actually zero) at the Equator and at a maximum at the poles. So what exactly does it do?

Let's imagine we fire a missile or rocket from the Equator, due north, that is towards the North Pole. To make the numbers easy, let's assume it leaves the launch pad at a velocity of 600 nm an hour and that it can maintain that speed indefinitely. This means that it can cover just about 10° of latitude in an hour. But when it is fired, both it and its launch pad are travelling eastwards at 901nm an hour – and in the absence of friction or some other resistance it keeps moving to the east at this rate as it flies north. After an hour it will have reached latitude 10° north. But we know from our calculations above that the easterly velocity of the surface of the earth at this latitude is not 901 nm an hour but only 888 nm an hour. So in the hour that our rocket has moved 901 nm to the east, a point on the earth's surface 10° due north of the launch pad will have moved only 888 nm. Consequently, after one hour our rocket is at 10° north, but is now 13 nm to the east of the direct line (of longitude) between the launch pad and the North Pole!

If the rocket keeps going at the same speed for another five hours it will reach a latitude of 60° north. But we know that at this latitude the earth is rotating at only half the speed of the equator. So while the rocket has moved eastwards 5,406 nm (6 x 901 nm), the earth's surface at 60° north of the launch pad has moved only half this distance, 2,703 nm. Now our rocket is no less than 2,703 nm to the east of its due north trajectory – and what's more, it is speeding eastward (at over 450 nm an hour) at three quarters as fast as it is flying north! Of course, after about 9 hours and flying 5,400 nm north the rocket will eventually pass over the North Pole and will be due north of its launch pad. But if you drew the line of its trajectory on the surface of a globe it would be a very strange curved line continually turning to the right. If you carried out the same experiment, but firing the rocket towards the South Pole, the result would be an exact mirror image, this time with the trajectory curving to the left.

At the other extreme, imagine standing precisely at the North Pole and firing the rocket to the south[10]. This time the rocket has no easterly velocity as it leaves the launch pad, but as it moves south, the earth beneath it moves more and more rapidly to the east (to the left of the rocket's course). The rocket's trajectory curves to the right *relative to the earth* just as it did when it was flying north from the Equator, but this time it appears to be moving towards the west. And whereas the eastward displacement between the Equator and 10° north was only 13 nm, in the hour that the rocket takes to cover the same latitudinal distance, that is 10°, but now from the North Pole to 80° north, its apparent displacement is no less than 156 nm. Now you can understand two important principles about Coriolis Force. First, that the path of anything moving over the earth's surface, but not coupled with it through friction, will tend to curve to the right in the northern hemisphere and to the left in the southern hemisphere *irrespective of the direction of its initial movement.* Second, the intensity of this tendency is at a maximum near the poles and a minimum, in fact zero, at the Equator.

And that takes us to the most famous myth about Coriolis Force, that as you cross from one side of the equator to the other, the way that water runs out of a plughole suddenly changes from anticlockwise[11] in the northern hemisphere to clockwise in the southern hemisphere. Many of us saw it demonstrated to Michael Palin with someone carrying a basin of water just a few metres from one side of the equator to the other and showing the water run out of the plughole in opposite directions on either side. Nice idea, but rubbish. We have seen that at the equator the effect of Coriolis Force is zero, so it couldn't possibly cause the demonstrated change in flow direction. In fact, even near the poles the force is probably insufficient at the scale of a normal washbasin[12] or bath to have any significant effect, so the whole argument is flawed in the first place. How, then, was Michael Palin fooled? By a simple, but very effective trick. At either end of his short walk across the equator, the demonstrator carrying the basin of water made a quick

---

10   Apart from vertically upwards, what other direction could you fire!

11   Anticlockwise, that is curling to the left because, the reasoning goes, currents of water in the basin or tub approaching the plughole move to its right-hand side and therefore go down the hole in an anticlockwise spiral.

12   However, some carefully conducted experiments in the 1960s, and reported in the scientific journal *Nature*, suggested that if great care was taken to minimise the effects of factors other than the earth's rotation, such as residual water movement from filling or using the bath or basin and temperature differences within the water itself, there was some indication that Coriolis Force did exert an influence at this scale. But under normal circumstances these external factors are much more powerful than the earth's rotation, so that there is no justification for the claim of a regular anticlockwise exit in the northern hemisphere and clockwise in the south.

little turn immediately before pulling the plug. He turned to his left at the northern end and to his right at the southern end. This was just enough to give the water a bit of anticlockwise momentum in the north and clockwise in the south. And this, in turn, was enough to ensure the direction of flow out of the plughole. Try it yourself; you can make the water flow in either direction wherever you are.

20 What causes ocean currents?

To answer this, let's look first at the major surface currents in the world ocean.

Every schoolchild learns about the Gulf Stream in the western Atlantic. This remarkable 'river in the sea', we are told, brings huge amounts of water from somewhere vaguely in the Caribbean past the eastern seaboard of the USA and then across the Atlantic to bathe British shores in warm water. This ensures that north-western Europe has a much more pleasant climate than places on the other side of the Atlantic at more or less the same latitude north of the equator.

This is mostly more or less true, but what is not often mentioned is that the Gulf Stream is just one part, albeit a quite remarkable part, of a global system of surface currents that transfer vast volumes of water, and enormous quantities of heat, over the surface of the earth and, in doing so, contribute hugely to our climate and weather.

The details of this current system are very complex, but the basics are rather simple (see Fig 7). In each of the two largest oceans, the Atlantic and the Pacific, there are two great circular current systems, called gyres, one in the northern hemisphere and one in the southern hemisphere. The two northern gyres flow in a clockwise direction, with the Gulf Stream in the Atlantic and a similar strong current, the Kuroshio, in the Pacific forming the western sides. The southern hemisphere gyres flow in the opposite direction, that is anticlockwise, though they are not quite as dramatic as the northern ones. Where these two great circular systems come together close to the equator there are two westward-flowing currents, the North and South Equatorial currents, separated by a relatively narrow eastward flowing Equatorial Counter-current.

A similar system of clockwise and anticlockwise gyres and an eastward flowing counter-current is found in the Indian Ocean but, because of the presence of Asia and the Indian subcontinent in the northern part, everything here is a bit squashed up and pushed further south than in the other oceans.

There are, of course, lots of ancillary and local currents, but the final part of the main current system is the West Wind Drift which flows in an easterly direction around the Antarctic Continent to the south of South America, Africa and Australasia.

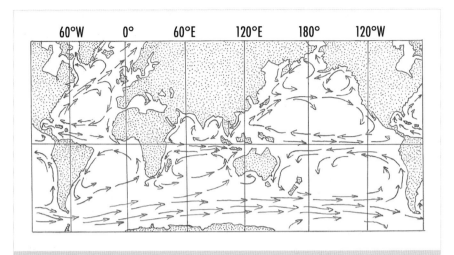

Figure 7
The major surface currents of the world's oceans.

That's the general current pattern[13], so what causes it? Basically, three factors: the wind, the rotation of the earth and the shape and position of the land masses.

The effect of the wind is most dramatically seen in the West Wind Drift since it occurs in one of the windiest parts of the earth, between about 40 and 65° south. These latitudes in the southern hemisphere are often referred to as the roaring forties, the furious fifties and the screaming sixties, and with some justification, because they are dominated by strong westerly or north-westerly winds producing an easterly flowing current[14]. Since there is no land mass in the path of this current it can flow continually round and round the earth.

If the earth was entirely covered by ocean, with no emergent land masses, there would be a whole series of such continuous currents flowing more or less parallel with the equator and driven by the prevailing winds. For example, at about 10–20° both south and north of the equator there would be westerly flowing currents driven respectively by the south-easterly and north-easterly trade winds (see Q98). And at around 40–60° north, the latitude of northern Europe, there would be an easterly flowing current, similar to the West Wind Drift, this time

13   Note that these current descriptions and maps are what is called 'climatological', that is based on averages of observations made over a long period. Consequently, on any particular occasion a ship may encounter a quite different current from that indicated on the chart. Mariners are well aware of this.
14   Somewhat confusingly, winds are always named with reference to the direction from which they come, while currents are named for the direction in which they go; so a north wind blows from the north to the south, whereas a northerly current flows from the south towards the north!

driven by the prevailing south-westerly winds in this region. Notice that these currents would not be flowing in exactly the same direction as the wind driving them. Instead, the currents would, and indeed do, flow at an angle of about 45° to the left of the wind direction in the southern hemisphere and to the right in the northern hemisphere. These deflections are a result of the rotation of the earth from west to east on its polar axis (see Q19).

But the earth isn't entirely covered by ocean. Instead, the northern hemisphere is dominated by land masses, North America, Europe and Asia, with huge terrestrial tongues extending into the southern hemisphere to separate off the Atlantic, Pacific and Indian Oceans. These land masses get in the way of the continuous currents and allow Coriolis Force, due to the rotation of the earth (see Q19), to come into its own. This causes the major currents in the northern hemisphere to turn to the right and those in the southern hemisphere to turn to the left. As a result, the westerly currents in the Atlantic and Pacific at 10-20° north, the North Equatorial Currents, both turn right (that is north) in the western extremity of their respective basins, producing the Gulf Stream in the Atlantic and the Kuroshio in the Pacific. These 'western boundary currents' then link up with the easterly flowing currents at 40-60° north. But when these currents reach the eastern side of their respective basins they are deflected to the right (that is to the south), to link up again with the North Equatorial Currents to complete the two great clockwise gyres.

Exactly the same happens in the southern hemisphere, but with the deflections to the left instead of to the right, resulting in the characteristic anticlockwise gyres of the southern Atlantic, Pacific and Indian Oceans.

And there, in essence, you have it!

21 Are there deep currents in the ocean?

There are, indeed, lots of them. In fact, the deep oceans are a bit like a layer cake, with a series of water masses moving in different directions at different levels.

The ocean's surface currents are driven by the winds (see Q20), and may flow quite rapidly, up to several knots. (One knot, by the way, is almost exactly half a metre per second.) In contrast, the deep currents are driven by density (or weight) differences between water masses and usually flow at no more than a few cm per second, that is a fraction of a knot. The density of seawater is determined mainly by its temperature and salinity, so that cold and very salty water is heavy and tends to sink, while warmer and less salty water is lighter and tends to rise. The deep current system is therefore known as the thermohaline circulation, *thermo* meaning heat and *haline* meaning salt.

The heavy and cold waters filling the deepest parts of the ocean sink from the surface at high latitudes, both in the Arctic and the Antarctic, to be replaced by warm surface currents (like the Gulf Stream) flowing from the tropics towards the poles. The resulting current system is extremely complex. But its main features have been summarised (some would say oversimplified) in a much reproduced illustration (see fig 8) of what is known as the great ocean conveyor or global conveyor belt because it transports huge amounts of heat around the globe and is largely responsible for the earth's major climatic zones. The conveyor idea, and its potential consequences, was first put forward by a highly respected American oceanographer, Wally Broeker, in the 1980s and was the basis of the 20th Century Fox apocalyptic science fiction film, *The Day After Tomorrow*, released in 2004.

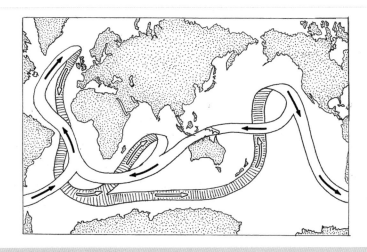

Figure 8
The global conveyor belt: a simplified version of the complex circulation of the oceans. The white parts of the conveyor indicate relatively warm and light surface waters and the cross-hatched parts represent cold, heavy deep waters.

The key features of the conveyor system are that the main source of deep oceanic water is the north-eastern Atlantic, where very cold and salty water from the Arctic ocean flows south-westerly across the relatively shallow sea floor between Scotland and Iceland (and particularly between the Faeroes and Shetland) and sinks beneath warmer waters brought across the north Atlantic in the Gulf Stream. This North Atlantic Deep Water then flows into the various deep basins of the Atlantic and some of it comes back to the surface in the Southern Ocean. But part of it continues around southern Africa contributing to the deep waters of both the Indian and Pacific Oceans. Some of this water finally surfaces in the north Pacific for the first time since sinking in the north Atlantic, after a transit time of 1,000 years or more. A complex warm water surface system of wind driven currents then brings water back to the Atlantic to begin the whole process again.

Quite apart from the significance of the conveyor in linking the major oceans together into a single unified system, it has attracted a great deal of interest in recent years because of its possible sensitivity to global climate change. Since the principal driver of the conveyor seems to be the sinking of cold, salt water in the North Atlantic, a number of oceanographers and climatologists have suggested that global warming might reduce or even stop it. The argument is that melting of the Greenland icecap and increased outflow of fresh water from North American rivers might lower the salinity of the North Atlantic and, along with a general warming of the oceans, could reduce its density to the point when it would no longer sink. If this happened, the argument goes, the Gulf Stream itself would greatly slow down or even stop. Paradoxically, then, global warming could lead to a significant cooling in north-eastern Europe, or even the onset of a new ice age!

There is good evidence that something of the sort has happened in the past, most recently in a period called the Younger Dryas beginning about 13,000 years ago. It lasted only about 1,300 years, but showed a rapid decrease in temperatures over wide areas of the northern hemisphere following a period of intense warming. The most likely explanation is the temporary cessation of the global conveyor system following a huge influx of fresh water into the Atlantic from Lake Agassiz, an immense glacial lake that occupied much of North America at that time.

This was the scenario explored in the film *The Day After Tomorrow*, though in this case extrapolated to rather Pythonesque extremes in which the northern United States turns into something akin to a deep freeze. The storyline was daft, but the science on which it was based, was good.

22 What are Tectonic Plates?

Tectonic plates are huge sections of the earth's crust (the so-called lithosphere) that are floating on a much denser layer, the asthenosphere, which consists largely of molten rock. There are half a dozen or so very large tectonic plates; the North and South American Plates, the Eurasian Plate, the African Plate, the Indian Plate, the Antarctic Plate and the Pacific Plate, along with a number of smaller plates filling gaps between the large ones (see Fig 9).

Some of the plates, such as the huge Pacific Plate and the smaller Nazca Plate off the western coast of South America, consist entirely of oceanic crust with no dry land apart from fairly small, and mostly volcanic, islands. Others, such as the Eurasian and North and South American Plates, include both oceanic crust and big

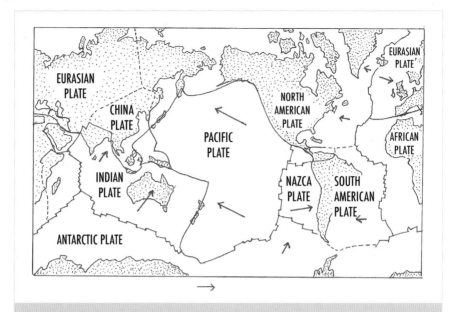

Figure 9
The earth's main tectonic plates. The direction and relative velocity of movement of the plates is indicated by the arrows. The plates in the Atlantic Ocean are moving at 1 to 2 cm a year, whereas those in the Pacific are moving two or three times as fast. The new seafloor contributing to these movements is produced mainly at the mid-ocean ridges, indicated on this diagram by the square zigzag lines running through the oceans.

lumps of rather lighter (less dense) continental crust. As its name suggests, continental crust forms the geology of the emergent continents (plains, deserts, mountains and so on) and also underlies the shallow continental shelf seas (see Q13).

New oceanic crust is produced at spreading axes, mainly on the mid-ocean ridges, where molten rock (magma) rises from the asthenosphere and moves away from the axis on both sides as it cools, contracts and sinks to form the new sea floor. This is the process of sea floor spreading or plate tectonics which, over many millions of years, has caused the continents to move over the earth's surface to their present positions. As it moves further and further from the axis the new material becomes covered in sediments, building up to layers kilometres thick in places (see Q17). Eventually, it descends back into the bowels of the earth, so to speak, when it is subducted, that is forced down beneath a neighbouring plate. The subduction zones are found either where the oceanic crust comes up against the edge of a piece of continental crust, such as the western coast of South America, or at the bottom of deep sea trenches (see Q16) such as those in the western Pacific.

Life In The Oceans

General

23 Are seaweeds plants?

Yes, and they need nutrients (phosphates, nitrates and so on) and sunlight to survive and grow just like the land plants we are much more familiar with. But whereas most terrestrial plants belong to a relatively advanced group, the flowering plants, which reproduce by growing from seeds produced within flowers, seaweeds – as their name suggests – belong to a much more primitive group, the algae or cryptogams, which reproduce with a much simpler system without flowers.

However, an important feature of both plant groups, and a crucial difference from animals, is that they can use the energy from sunlight to make complex chemical molecules from simple salts and carbon dioxide in the process called photosynthesis. But while most flowering plants obtain their carbon dioxide from the air surrounding them and their nutrients from the soil via their roots, seaweeds get all their chemical supplies from the seawater and not from the underlying sand, mud or rocks. Consequently, seaweed 'roots' have a quite different function. In fact, they are called holdfasts, not roots, and the name reflects their role. The function of the holdfast is simply to anchor the seaweed in a suitable place and to stop it being washed around by the waves and currents.

Just as is true for terrestrial plants, 'a suitable place' will differ for each algal species, but an essential requirement for all of them is that they must be anchored where there is sufficient sunlight for photosynthesis. This represents only a very tiny proportion of the sea floor and severely limits the distribution and significance of fixed seaweeds in the economy of the oceans (see Q24).

24 How deep do seaweeds grow in the ocean?

About 100 metres (or 300 feet).

Seaweeds, like all plants, need a minimum amount of light to grow; in fact, about 10 watts per square metre (10 Wm^{-2}). This is about one one hundredth (or 1%) of the intensity of bright sunlight at the surface (which is about 1,000 Wm^{-2}). It is the depth at which the gains from photosynthesis are exactly balanced by the losses from respiration, and is known as the compensation depth.

As sunlight passes down into the sea its intensity is reduced, by both absorption and reflection, and the more particles there are in the water, the more rapid this reduction. Even in the very clearest oceanic waters, containing almost no suspended particles (see Q8), the intensity falls to 1% of the surface value at a depth of about 150 metres. However, apart from those that float at the surface, seaweeds don't live in the open ocean. Instead, they live anchored to the bottom near the shore. For a whole variety of reasons, inshore waters have many more suspended particles than oceanic waters, so the rate at which the light intensity decreases with depth is much greater.

In some special areas the water is clear enough to allow effective photosynthesis as deep as 40 or 50 metres (150 feet or so), but in most places it is no more than 10 or 20 metres (50 or 60 feet) at most and often even less. Consequently, the seaweeds with which we are most familiar are the green, red and brown ones we see attached to rocks between tide marks and the deep 'rooted' large brown kelp whose broad fronds float at the surface at low tide[15].

Fascinating though these shallow living 'rooted' algae are, they are relatively unimportant in the economy of the seas generally. Only a very tiny proportion of the oceans are shallow enough for sufficient sunlight to reach the bottom and therefore to support the growth of anchored seaweeds. The animal life in the rest of the ocean, something like 98% of its total surface area, is dependent on a quite different group of plants, the planktonic algae or phytoplankton, of which most of us are totally unaware (see Q27). But just like the seaweeds, these planktonic plants are also dependent upon sunlight to survive and grow. So although the odd planktonic algal cells might just about get by at a depth of 150 metres in the deep tropical oceans, their high latitude cousins living in much more productive

---

15   For more than 100 years these large kelp have been known to be very efficient accumulators of iodine. Recent research indicates that, when stressed (for example exposed to the air by very low tides), they release some of this iodine into the atmosphere and possibly contribute to cloud formation and therefore affect the local climate. Who would have guessed that?

and therefore murkier waters, particularly in coastal regions, cannot afford to be more than 10 or 20 metres from the surface if they are to receive enough light to photosynthesize successfully. A fascinating corollary of this simple fact is that virtually all of the myriads of animals living in the oceans, all the way down to the bottoms of the deep trenches, are absolutely dependent for their existence on the plant growth in this paper-thin surface layer.

25 Why are seaweeds mainly brown?

We are so used to terrestrial plants being green that it can come as something of a surprise to see marine plants, like the familiar seaweeds on rocky shores, that are anything but green. Of course, there are some green seaweeds, and also quite a few red ones, especially among the smaller varieties, but the majority of inter-tidal and subtidal seaweeds, the wracks and kelps, are, indeed, mainly a yellowish-brown. So why?

Plants perform a crucial function for all animal life on earth, including human life, because they can do something that no animals can. They can use the energy in sunlight, in the process of photosynthesis, to produce complex molecules, which we animals need, from simple molecules like carbon dioxide and water. A pivotal role in this process is played by photosynthetic pigments that trap the sun's energy within the plant cell and make it available for the complex chemical reactions. The principal photosynthetic pigment in terrestrial plants is chlorophyll which is green and gives plant leaves their characteristic green colour. Chlorophyll looks green because it absorbs the long wavelength (red) and short wavelength (blue) parts of the visible spectrum (using these to provide the plant with energy) and reflects, that is essentially rejects, the green light more or less in the middle of the spectrum.

However, when sunlight enters the sea its wavelength make up (that is its colour) is changed dramatically with depth (see also Q8). In summary, the shorter (and more blue) the wavelength, the further the light penetrates into the sea. Consequently, the longer wavelengths (that is the red end of the spectrum) disappear first, and the shorter wavelength blue light travels furthest. So chlorophyll is a fairly good pigment for a marine plant to have if it lives very close to the surface (where there is still plenty of both blue and red light), and quite a number of intertidal seaweeds living high on the shore, and therefore never covered by deep water, are, indeed, green. Chlorophyll is also found in lots of the microscopic phytoplanktonic algae living in the near surface layers (see Q27). But for a seaweed to live much deeper under the water it needs a photosynthetic pigment that is more efficient than chloro-phyll at absorbing the more and more blue light that reaches it. In fact, all seaweeds contain chlorophyll but its green colour is masked by the other pigments they contain. Red seaweeds seem to be particularly efficient in very turbid waters, but in deep and fairly clear waters the brown algae come into their own. Their principal pigment, fucoxanthin, absorbs blue-green light particularly efficiently, but reflects all the other parts of the visible spectrum, giving them their yellowish-brown appearance.

26 What is plankton?

The term plankton is derived from a Greek word meaning to wander or drift. It is used by biologists to describe all those animals and plants (in both fresh water and the sea) that live in midwater and, unlike many fish, whales, seals and so on, are unable to swim against the currents; instead, they are simply carried along by them. So 'plankton' includes lots of different species belonging to many different groups, some of them living the whole of their life cycle in this way, and others spending only their young stages in the plankton.

All planktonic plants (the phytoplankton, see Q27) are tiny, and so are the vast majority of planktonic animals (the zooplankton). But some planktonic animals, like the jellyfish (see Q33), can be quite big, since the only qualification for inclusion in the plankton is being unable to swim strongly enough to resist being carried along by the water currents.

However, although planktonic animals can't significantly control their horizontal movements, many of them are capable of swimming vertically through considerable distances. Some of them regularly move upwards and downwards through several hundreds of metres every single day. This is part of a remarkable worldwide phenomenon called 'diurnal vertical migration' in which billions of tonnes of planktonic animals regularly move up towards the surface as the sun goes down in the late afternoon and evening and then make the return journey to their daytime depth as the sun rises again at dawn. This regular movement is also followed, of course, by many non-planktonic animals that are dependent upon the plankton for food.

27 What is phytoplankton?

The phytoplankton is the basis for virtually all life in the oceans, from the tiniest invertebrates to the largest whales, and from the surface waters to the bottom of the deepest trenches. The term comes from the Greek words *phyton* meaning plant and *plangtos* meaning wandering, and refers to all the plants drifting in midwater both in fresh water and in the oceans. They are all single-celled organisms, though they sometimes stay joined together in quite long chains. All of them are very small, even the biggest individual cells reaching no more than 1 mm across. At the other extreme, the smallest, called the picoplankton, can measure as little as 0.2 microns, so that it would take 5,000 of them laid side by side to stretch to one millimetre!

The importance of the phytoplankton, compared with their much larger and more obvious plant cousins the fixed algae (usually called seaweeds, see Q23), has been recognised for well over a hundred years. For most of this period the larger phytoplankton cells were thought to be the main producers, but modern research suggests that, though the larger cells are certainly quite significant, the real power-house of oceanic plant production is at the smaller end of the size range.

As plants, phytoplankton are autotrophs (self-feeding), that is, they make complex compounds out of simple ones like carbon dioxide and water using the energy from sunlight in the process of photosynthesis, though some of them manage to blur the boundaries between plants and animals.

The phytoplankton includes members of many different plant groups, but the main ones are diatoms, coccolithophores, dinoflagellates and cyanobacteria.

Diatoms include the largest phytoplankton cells. They are enclosed within glass-like cases made of silica often in beautiful geometric shapes. After the diatom's death these cases, called frustules, sink to the seabed and may be so abundant that they dominate the sediments forming a glassy diatomaceous ooze.

In contrast, the cells of coccolithophores are enclosed in 'shells' consisting of a series of tiny overlapping circular plates, called coccoliths, made of calcium carbonate or chalk. Coccolithophores can occur in such vast numbers that when they die and release the coccoliths they colour the surface waters a milky white that can be seen from space. This increases the reflectivity of the sea surface so that less heat from the sun is absorbed. Like the diatoms, the remains of dead coccolithophores can form deep sediments, and the famous white cliffs of Dover on the coast of Kent, south-east England, are composed almost entirely of coccoliths deposited when this region was beneath tropical seas more than 130 million years ago.

Dinoflagellates are strange little organisms that are more mobile than the diatoms and coccolithophores because they can row themselves about to some extent using two tiny whips or flagella, hence their name. They include many species that are autotrophs, and therefore regular plants, but others eat other organisms, just like animals. As if to confuse matters a bit more, some dinoflagellates seem to be able to live both as plants and animals! Among their many interesting attributes, many dinoflagellates are able to produce living light or bioluminescence (see Q37) and may be responsible for spectacular displays when they occur in large numbers. At the same time, several dinoflagellate species produce rather nasty toxins and may sometimes occur in huge numbers and pose a serious threat to other oceanic organisms.

Finally, the cyanobacteria, also known as blue-green algae, are photosynthesising bacteria that occur in almost every conceivable environment from desert sand, bare rock, soil, in partnership with other organisms and, of course, in the ocean. They seem to have been around for a very long time, approaching 3 billion of the 4.5 billion years of the earth's existence. In fact, the ability of some early forms to produce oxygen more or less from scratch, so to speak, seems to have changed the planet's atmosphere dramatically and allowed the explosion of biological diversity that we see today. A downside of cyanobacteria is that, like the dinoflagellates, several of them produce toxins and can also occur in dangerously high concentrations.

Like all plants, phytoplankton cells are absolutely dependent upon sufficient sunlight for photosynthesis. Consequently, they can grow only down to depths of 100 metres or so in the clearest mid-oceanic waters and to much shallower depths, often only a few metres, in turbid inshore waters. In fact, in many shallow water areas phytoplankton cells themselves contribute to the turbidity and colour of the water. So the green, yellow or brownish colour of such waters, often attributed to pollution, may actually be an indication of a particularly rich and healthy ecosystem. This is certainly true of the North Sea which, despite being subjected to considerable pollution from the surrounding human populations, is an extremely productive piece of ocean.

Between them, the total annual growth (called primary production) of marine phytoplankton is estimated at just about 50 gigatonnes of carbon (that is 50,000,000,000 tonnes!). Terrestrial plants are estimated to produce a little bit more than this, but since the oceans cover about twice the surface area of the emergent land this means that, on average, a square metre of ocean is less than half as productive as the land. This is because the oceans, like the land, have highly productive areas and also very barren 'desert' areas, with the oceans' deserts being rather larger than the land's. Paradoxically, while the land's most productive regions tend to be in the tropics (along with some of the deserts), the oceans' most productive areas are in the colder high latitude regions (see Q28).

28 Why are some parts of the ocean more productive than others?

The productivity of the oceans, whether you measure it by the amount of fish and shellfish that we take out of it, or by the number of mammals and birds you see as you sail across it, is ultimately dependent upon the amount of plant growth going on in the surface waters, particularly by the phytoplankton (see

Q27). Plant growth in the sea, as on land, is all down to sunlight, nutrients and temperature, but in the ocean the way these work is possibly not quite as you might expect.

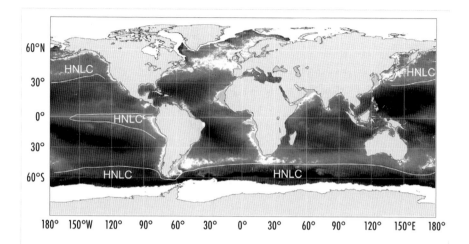

Figure 10
Productivity of the oceans based on the amount of chlorophyll in the surface water measured by the SeaWIFS satellite in 1998 and 1999. The white areas adjacent to the poles represent sea ice. Otherwise the white and lighter grey areas indicate high chlorophyll levels while the darker areas indicate lower chlorophyll levels. (Reproduced courtesy of Dr Mike Whitfield and Ocean Challenge, the journal of the Challenger Society for Marine Science

A good indication of productivity is the amount of chlorophyll in the water, and these days this can be measured from space. Fig 10 summarises the average levels of chlorophyll in different parts of the oceans in 1998 and 1999 detected by a specialised satellite called SeaWIFS[16]. This black and white version was produced by Dr Mike Whitfield, while a rather prettier coloured version of a longer data set is reproduced inside the back cover. Both versions show a number of interesting features. First you will notice that a narrow strip around all of the earth's main landmasses is pretty productive. These are the waters over the Continental Shelves, the shallow areas extending from the coastline to the continental edge at a depth usually of about 200 metres. Q13 explained that the British Isles stand entirely on

---

16 SeaWIFS stands for Sea-viewing Wide Field-of-view Sensor, not exactly the catchiest title, but a very clever satellite specifically designed to measure ocean characteristics like chlorophyll and water clarity. It has been collecting data since September 1997.

the Continental Shelf off north-western Europe so that both the North Sea and the Irish Sea are shallow water seas and, from Fig 10 and the cover picture, you can see that they are very productive.

There are good reasons why continental shelf waters are generally rather productive. First, it is into these inshore waters that the world's rivers drain, carrying a chemical cocktail that includes the nutrients that the marine plants need. But continental shelf waters are also, by their very nature, rather shallow, rarely more than 200 metres deep and usually much less. The average depth of the North Sea, for example, is only about 90 metres and much of it is a great deal shallower than this. So there is an absolute limit to how far anything can sink in continental shelf waters before the bottom intervenes. This applies to individual living or dead plankton cells, of course, but also to the nutrients or any particles that are containing them like faecal material or the dead remains of animals and plants. Because they haven't been able to sink very far, these bits and pieces, including the nutrients, can be brought back to the surface sunlit layer by water currents and storms that stir up the water column. So phytoplankton over the continental shelves are rarely short of nutrients. Their problems, if they have them, are much more likely to be getting sufficient sunlight to allow photosynthesis to take place, particularly during the winter months in high latitudes and at more or less any time if there is too much sediment suspended in the water.

Now look at the central oceanic regions in Fig 10 and the inside cover picture, and you will notice some remarkable things. You will see that the very lowest chlorophyll levels are found in the subtropical regions of each of the oceans, the areas identified as oceanic gyres in Q17 surrounded by currents flowing in a clock-wise direction in the northern hemisphere and anticlockwise in the southern hemisphere. These are the true deserts of the ocean. If you sail through them you will see lots of brilliant blue water (because there is nothing in it, (see Q8)) quite a few flying fish and the odd migrating whale, dolphin or turtle. But you will emphatically not see large numbers of seabirds, nor will you see lots of fishing vessels. The reason is obvious, it is because there is not enough life for the seabirds to feed on and certainly not enough fish to support big fishing fleets; but why? Well it is all down to water temperature.

Q18 pointed out that superheated seawater at temperatures up to 300-400°C gushes out of the seabed at some hydrothermal vents. But this truly remarkable phenomenon is extremely local and, within a few metres of the openings of the hottest vents, the water temperature drops to no more than 2-3°C. In the rest of the oceans the total temperature range is from about 40°C in the warmest trop-ical lagoons to about minus 1.9°C in the polar regions. But if you could take a giant whisk and mix all the oceanic waters together the average temperature would be around 3-4°C, that is about the temperature of the inside of the average domestic fridge. This is because the bulk of the oceans are pretty cold. At a depth of 1,000 metres under the tropical midday sun the temperature will be no more than 5°C while the vast majority of the deeper parts of the oceans are filled with water between about 0° and 2°C. At the other end of the conventional temperature scale, water temperatures in excess of about 20°C are restricted to a very thin surface layer in the tropics and subtropics, say between about 35° north and 35° south. With warm water at the surface, and cold water deeper down, there is inevitably somewhere in between where the temperature changes quite dramatically. This is called a permanent thermocline and, because cold water is much denser (heavier) than warm water, it is a stable situation which effectively stops the water masses on either side of the thermocline from mingling. More importantly, although lots of midwater animals can, and do, swim up and down through the thermocline, anything that sinks through it, like animal carcasses or dead plant cells and all the nutrients locked up in them, are essentially lost to the surface layers unless and until special conditions such as upwelling (see later) bring them back to the surface. And this is the key to variations of productivity at different latitudes.

Apart from the very high latitude polar seas, where the water temperature can be pretty cold from top to bottom, virtually all areas of the ocean have a permanent thermocline, but its depth and efficiency as a barrier varies enormously. At mid latitudes, what you might call temperate seas, say from about 35° to 60° both north and south of the equator, the surface water temperature during the winter is generally between about 6°C and 15°C. Under these circumstances, winter storms mix up the surface layer to a depth of several hundreds of metres in the open ocean and right to the bottom over the continental shelves. Moreover, there is some mixing through the rather weak thermocline down to about 1,000 metres depth, so some of the sinking material will be returned from these depths back to the surface. Things change dramatically in temperate seas during the spring and summer, but before looking at that let's see what happens in the tropical and subtropical seas.

Here, in both summer and winter there is a lot of incoming heat from the sun and the surface waters stay warm all year round, mostly well over 20°C. The mixed layer is very thin, usually no more than 100 metres or so. Below this depth the water temperature decreases rapidly, dropping 10°C over no more than a few tens of metres. This thermocline is extremely strong, far too strong to be disturbed by winter storms. Consequently, nutrients that sink below 150-200 metres are effectively lost. As a result, subtropical oceanic surface waters have plenty of sunlight and plenty of warmth, but they are almost always short of the third necessary ingredient for plant growth – nutrients, so they produce little phytoplankton or anything else!

Now let's go back to those temperate waters. We left them during the winter with storms stirring up the surface layer down to a depth of several hundreds of metres, ensuring that lots of nutrients are being brought back to the surface. However, an inescapable consequence of this mixing is that the over-wintering phytoplankton cells, what you might think of as the seed corn of the next year's crop, are also mixed throughout this layer. So they spend a lot of time well below the depth at which sufficient sunlight for photosynthesis would penetrate during the summer, let alone during the much duller winter months. As a result, despite being surrounded by loads of nutrients, no significant phytoplankton growth takes place; just like on land, the oceans are in a dormant, over-wintering state. Then, in the spring the situation changes almost magically.

As the sun's strength improves in the spring (March and April in the northern hemisphere, September and October in the southern hemisphere), it begins to warm the surface layers of the ocean. Initially, this is by only a few tenths of a degree and the effect is rapidly negated by storms mixing the slightly warmer water with the colder layers beneath it. But eventually, given a few days of reasonably calm weather, an embryonic thermocline becomes established a few tens of metres beneath the surface. Once the temperature change at this new thermocline reaches half a degree or so it becomes quite difficult for all but the most severe storms to disrupt and the process of warming up the surface layers, isolated from the deeper colder water by this seasonal thermocline, can continue.

However, while the increase in temperature of the surface waters helps algal growth just as the warmth of spring encourages land plants, a much more important consequence in the sea is the thermocline acting as a barrier to mixing between the surface and deeper layers. Any plant cells caught beneath the thermocline once it is established are doomed. They might be surrounded by plenty of nutrients, but they are far too deep to receive consistently enough sunlight to be able to photosynthesise. Consequently, they will either be eaten by a tiny midwater herbivore or, after a few days of decline, they will die and sink to the seafloor to be eaten by a detritivore. Algal cells above the thermocline, on the other hand, have hit the jackpot. For the first time since the previous autumn they are in a situation where they will never be carried down below the seasonal thermocline; and as long as this is no deeper than the compensation depth (see Q24) they will always have enough sunlight during the day to photosynthesise, grow and reproduce. And before you know it, you've got a spring algal bloom to feed all the little midwater herbivores which, in turn, feed the little carnivores, then the middle sized carnivores and then the biggest carnivores – including us!

So this, in essence, is why the mid to high latitudes are the breadbaskets of the oceans while the tropical and subtropical gyres are the deserts. But before leaving this topic, there is one other very important phenomenon to take into account – upwelling. If you look back at Fig 10 and the inside cover picture you will see some strange stripes of high chlorophyll (i.e. productivity) stretching away from continental coastlines out towards the middle of the tropical and subtropical oceans. There are two particularly striking thin bands crossing the tropical Atlantic from the Congo-Angola region of West Africa and the Pacific from the coast of Colombia-Peru, but there are also a number of less extensive high chlorophyll areas including the north-west African Coast, southern California, the coast of the Horn of Africa and Saudi Arabia, and off New Zealand. In all these areas a combination of the local wind conditions and the effect of Coriolis Force (see Q19), produces a situation in which nutrient-poor surface water is driven away from the shoreline and is replaced by cooler, nutrient-rich water upwelling from the deeper layers. Since most of these regions are permanently warm and amply supplied with sunlight, more or less year round, the effect of the injection of nutrients is dramatic and several of these regions support important fisheries.

Finally, we have to add an up-to-date twist to this story. Over the last 25 years or so, marine scientists of many nations have collaborated in studying the productivity of the oceans, and particularly the distribution of nutrients, in much more detail than ever before. The outcome of all this work broadly confirms the above explanation of productivity variations based largely on the physical characteristics of the water column. But it has also revealed some fascinating discrepancies. For while there is a strong correlation between the amount of nutrients available and the level of phytoplankton growth in most parts of the ocean, there are some large areas where this is not true. In these so-called 'High nutrient, low chlorophyll' areas (labelled HNLC in Fig 10), the amount of plant growth seems to be much lower than you would expect from the available nutrient levels. So what could be restricting the phytoplankton growth? In a word, the answer seems to be iron! For various, often complicated, reasons, the HNLC areas, especially the north-eastern and eastern-central Pacific and a good deal of the circum-Antarctic Southern Ocean do not receive enough iron to enable their phytoplankton cells to use the available nutrients efficiently.

This discovery sparked huge interest within the international marine scientific community, particularly because it pretty well coincided with the growth in concern about global warming and the role of atmospheric CO_2. Phytoplankton, like all plants, use carbon dioxide. Indeed, lots of atmospheric carbon dioxide is routinely taken up by the normal growth of oceanic phytoplankton and eventually becomes locked up, or sequestered, in the sediments of the deep sea floor when the plant cells and the animals that eat them die and sink to the bottom. However, we all know that the level of carbon dioxide in the atmosphere is inexorably rising and, among other things, is increasing the acidity of the oceans (see Q114). So lots of politicians, engineers and scientists around the world are beavering away to study ways of reducing the amount of new man-produced CO_2 entering the atmosphere and, at the same time, ways of removing some of the CO_2 that is already there. 'Ah,' thought some marine scientists, 'so what if we artificially make up for the lack of iron in the HNLC areas by spraying them with iron filings or some such? Would that stimulate the phytoplankton growth, which would then suck in the excess atmospheric carbon dioxide and eventually deposit it safely in the deep sea sediments where we could forget about it? If so, it might provide at least a partial solution to the global warming dilemma.'

Not surprisingly, this kicked off a whole plethora of experiments involving chucking the odd few tonnes of iron salts into various carefully selected bits of the world ocean and then seeing what happened. Sure enough, the first experiments, carried out in the HNLC region of the eastern central Pacific in 1993, definitely increased the phytoplankton growth. But disappointingly, it didn't result in increased carbon export to the deep ocean. Instead, the 'new' production stayed in the surface layers where it was recycled. So would iron fertilization help get rid of atmospheric CO_2 or wouldn't it? Nobody knew, and much more work was clearly needed, not least of all to make sure there would not be any unforeseen environmentally damaging side effects of any large-scale enrichments.

Several more enrichment experiments have been conducted in the sub-Arctic Pacific and in the Southern Ocean and more are planned. Indeed, in December 2008, under pressure from environmental groups, the German government halted a cruise to the Southern Ocean of one of its major research vessels to conduct an enrichment experiment because it was worried that it might contravene the UNCLOS conventions (see Q118). After a couple of days' discussion and consultation, the government relented and the cruise went ahead. But despite the original optimism, the results are likely to be interesting but disappointing.

All of the experiments so far conducted support the original ones; the addition of iron certainly increases the phytoplankton production, sometimes dramatically so. But the extra carbon dioxide taken up in the process does not seem to be deposited on the deep sea floor. And as I write this piece in March 2009 a paper published in the prestigious scientific journal *Nature*, describing a multi-national study in the area of the Crozet Islands, some 1,200 nm to the south-east of Cape Town, seems to confirm this conclusion. This detailed study found that the fallout of material to the sea floor from areas with a natural supply of iron was much greater than from areas artificially fertilized. The reasons are not clear, but it seems that much of the artificially supplied iron is 'lost' in various ways and is therefore not used efficiently by the phytoplankton. Sadly, the early hopes that iron fertilization might provide an almost miraculous solution to the CO_2 problem seem unlikely to be fulfilled. But the final verdict has yet to be delivered and the jury is definitely still out. Watch this space!

29 How deep in the oceans do animals live?

All the way down to the bottoms of the deep ocean trenches (Q16) at a depth of almost 11 km or 7 miles, and in the water column all the way from the surface to the bottom.

We know this, first of all because fish have been seen at the very bottom of the ocean (see Q115), but also because animals of all the main groups, from protozoans to vertebrates, have been collected from the deep sea floor with trawls, dredges and corers and from midwater with plankton nets.

However, with increasing depth in the oceans the total amount of animal life undoubtedly decreases very significantly, so that the amount on the bottom of a deep sea trench is no more than 100th of that found on an equivalent area of shallow continental shelf (Q13) waters. And the reason is – food! Life in the whole of the deep oceans is dependent for its existence on plant growth in the surface layers where photosynthesis can take place (see Q27). Most of this plant material is eaten close to the surface where it is produced, but some of it sinks to the deeper layers either directly, as dead or dying phytoplankton cells, or indirectly as the faecal pellets or bodies of the animals that fed on the phytoplankton. But this sinking material has to run the gauntlet of the hungry midwater creatures. Consequently, less and less food reaches the deeper layers so fewer and fewer animals can make a living. But even if there is only the slimmest chance of surviving in a particular environment, you can bet your life that some beast or other will have evolved to do so. And they certainly have in the deep ocean. Not only are animals found in the very depths of the ocean, but evolution seems to have made up for the limited total abundance of life in these inhospitable regions by furnishing them with a bewildering variety of different forms, possibly more than in any other environment on earth (see Q31).

30 Why aren't deep sea animals crushed by the pressure?

Q7 explained that the pressure in the sea increases by about one atmosphere or 15 pounds per square inch for every 10 metres of depth. So at 1,000 metres depth the pressure is equivalent to about 100 atmospheres, while at 10,000 metres down it is a staggering 1,000 atmospheres, that is 15,000 pounds per square inch or one metric tonne per square cm. Yet all sorts of animals, ranging from relatively simple sponges and sea anemones to much more complex fish, live quite happily under these seemingly horrific conditions. How on earth do their fragile bodies withstand the enormous pressures?

Well actually, they are no more conscious of the pressure they are living under than we humans are aware of the fact that we live at a pressure of about 15 pounds per square inch, or a total of between 15 and 20 tons theoretically pressing on the surface of the average adult. And the reason we don't feel it is simply because the pressure inside our bodies, especially the compressible air-filled spaces such as our lungs and our middle ears, is exactly the same as that surrounding us. Consequently, we become conscious of the pressure only if it changes rapidly and these spaces either expand or contract as a result. In these situations, for example climbing or descending rapidly in an aeroplane, or even going up or down a steep hill in a car, we may experience pain in our ears because of unequal pressures on either side of our eardrums. This is usually because our Eustachian tubes, the narrow canals linking our middle ears (the bits under that lump behind our ear drums) to our pharynx, are a bit bunged up. The discomfort is normally easily alleviated by either waggling your jaw from side to side or pinching your nose and trying to breathe out through it. Both procedures tend to dislodge the minor blockage and you can feel your ears pop as the pressure on both sides equalises.

In exactly the same way, most animals living in the sea don't experience sudden changes in pressure, and few of them would notice even if they did – for two very good reasons. First, the vast majority of them, unlike us, do not have compressible gas-filled spaces inside them like lungs and ears. Instead, their bodies, like the bulk of ours, is made up of cells, tissues and organs that are composed of liquids, mostly water. These substances are almost incompressible, that is, they hardly change their volume under different pressures and simply adopt the pressure of the surrounding water, retaining the same form, neither expanding nor contracting, as the animal moves up and down in the sea.

But what about the marine animals that do have gas-filled spaces, like the fish with their swim bladders?

Many bony fish, but by no means all, have gas-filled bladders inside their body cavities. They can fulfil various functions, but their main purpose is to act as a buoyancy aid to allow the fish to maintain a neutral buoyancy, that is with no tendency to either sink or float. For this to work, the swim bladder must retain the same volume at any depth the fish might live at. But if the fish swims upwards, into decreasing pressure, the swim bladder will tend to swell and the fish will become more and more buoyant. To counteract this, the fish has a mechanism for absorbing gas from the bladder back into the bloodstream. In the same way, if it swims deeper so that the increasing pressure tends to squeeze the gas bladder to a smaller volume, the fish will secrete more gas *into* the bladder to maintain its volume. In most circumstances these gas absorbing and secreting mechanisms can cope easily with the pressure changes experienced in their normal vertical movements. Paradoxically, however, this is much more difficult for a shallow living fish than for its deep living cousin! Here's why! Consider a fish swimming upwards sufficiently for its swim bladder to double in volume if its gas-absorbing mechanism didn't work. For this to happen, the outside pressure would have to be halved. A deep sea grenadier living near the bottom of the deepest ocean, say at a depth of 10,000 metres, is living at a pressure of about 1,000 atmospheres, so to experience a halving of the pressure it would have to swim halfway to the surface, to a depth of about 5,000 metres where the pressure would be about 500 atmospheres. In contrast, a herring living at 10 metres depth would be experiencing a pressure of just about 2 atmospheres. If it swam the 10 metres to the surface, the pressure would be reduced to one atmosphere and its gas bladder would have shown exactly the same doubling of volume as the grenadier's after swimming upwards 500 times further! Amazing or what!

31 How many different types of animals live in the ocean?

Nobody knows, but most deep sea biologists believe that the deep oceans are home to many more species than the 1.8 million or so described to date from all environments put together, possibly as many as tens of millions, most of them living in, on or very close to the deep sea floor!

This represents a real turnaround because if these same scientists (and I'm one of them) had been alive 150 years ago, they would almost certainly have believed that the oceans below a depth of a few hundreds of metres were totally devoid of any life at all. How this remarkable change of view has taken place deserves a bit of explanation.

Powered winches were rare on ships in the first half of the nineteenth century and almost everything was done by hand. Consequently, dredging or trawling in deep water was extremely time consuming and very hard work. In any case, the chance of catching anything seemed pretty remote, so few ship's officers or scientists thought the effort was worthwhile. After all, it was well known that light levels rapidly decrease with depth in the sea so that the bulk of the deep ocean must be absolutely pitch black. The importance of light for plant growth was also well known, but there was very little knowledge of the phytoplankton cells that we now know are crucial to the economy of the oceans (see Q27). Instead, scientists generally assumed that any food entering the deep sea would have to come from the shallow areas where seaweeds grow. Although nobody at that time knew exactly how deep the oceans were, it was well understood that pressure within them increased rapidly with depth (see Q7), so that any creature living deep down would surely be subjected to crushing pressures. Finally, it was known that the ocean depths were cold, though for most of the nineteenth century marine scientists didn't realise quite how cold, believing, quite wrongly, that they were filled with water at 4°C just like the bottom of a garden pond in winter.

With this awesome combination of total darkness, low temperature, huge pressure and no food it seemed very unlikely that anything could live in the depths of the ocean. So when a young medical student dropout from the University of Edinburgh got a job as a naturalist on a Royal Naval ship, HMS *Beacon,* surveying in the Mediterranean in 1841, his dredging results in the Aegean Sea seemed to confirm this view.

Edward Forbes had already dredged extensively in the relatively shallow waters around his native Isle of Man and, further afield, around Orkney and Shetland and was familiar with the rich fauna to be found on the sea floor in these regions down to depths of around 100 fathoms (180 metres). So when he was able to dredge down to 230 fathoms (420 metres) in the Aegean, he was intrigued to find that his catches decreased dramatically with increasing depth, both in the number of specimens collected and in the number of species represented. His results, concluded Forbes, pointed '. . . to a zero in the distribution of animal life yet unvisited'. In 1843 he produced his 'Azoic theory' according to which, as suspected, the deep ocean beyond a depth of a few hundreds of metres would be totally lifeless. By this time, although he was only 28 years old, Forbes was one of the stars in the Victorian scientific firmament and could have anticipated a glittering career comparable with those of Charles Darwin and Thomas Henry Huxley. Sadly, this was not to be, for he died in 1854 when he was only 39. Had he lived he would have been delighted to see his theory totally overturned in the 1860s and 1870s.

Forbes' misfortune was that he had based his ideas on results from the Aegean which, like the rest of the Mediterranean, is a very poor sea. Although animals are to be found at the very bottom of the Mediterranean at depths of 5,000 metres or more, the drop-off in abundance and diversity in this sea is much more dramatic than almost anywhere else on earth. But neither Forbes, nor his contemporaries, had any inkling of this and his Azoic theory was therefore widely accepted both in Britain and abroad for almost 30 years.

However, as steamships and steam winches became more common in the 1850s and 1860s, a number of marine scientists began to question Forbes' ideas and became intrigued by other aspects of the deep oceans, such as currents and seabed geology. Eventually, the gathering pressures led to a famous circumnavigation in a Royal Naval ship, HMS *Challenger,* from 1872 to 1876, during which biological, chemical, physical and geological samples and data were collected throughout the oceans, essentially establishing the modern discipline of oceanography. Among the vast amount of information gathered by the *Challenger* scientists, one conclusion stood out above all others. Their dredging and trawling work had demonstrated that, contrary to Edward Forbes' forecast, animals were to be found to the very greatest depth they sampled, at about 5,700 metres on the edge of what we now know as the Japan Trench in the western Pacific and not far from the deepest part of the world ocean, still carrying the name *Challenger* (see Q6). But it appeared that Forbes had been half right, because the *Challenger* results seemed to show that both the total abundance of animals and the variety decreased with increasing depth. In other words, in modern parlance, the deep sea was a low biodiversity environment.

This view remained pretty well unchallenged for the next 80 years or so, and in the 1950s it was still confidently believed that with increasing depth in the sea there were fewer and fewer animals belonging to fewer and fewer species. Between the 1880s and the 1950s there had been lots of major scientific expeditions from many countries following in the steps of the *Challenger,* but although they made huge improvements to our knowledge of the physics and chemistry of the oceans, they added relatively little to an understanding of deep sea biology. After all, although the *Challenger* scientists had not managed to trawl at the bottom of the very deepest parts of the oceans, their deepest trawl had brought up animals from deeper than about 96% of the deep sea floor. The presence of animal life in the remaining 4% was confirmed by trawling and, finally, by eye when Jacques Piccard and Don Walsh dived to the bottom of the Marianas Trench in January 1960 (see Q115).

But quite remarkably, over the whole of this period the technology for sampling the deep sea had not significantly improved since the days of the *Challenger.* So the plankton nets, trawls and dredges being used by deep sea biologists in the 1950s would have been totally recognisable to reincarnated *Challenger* scientists – and so would their catches! Unfortunately, these nets, and particularly the trawls and dredges dragged across the sea floor, gave a very biased view of what lives there. They were reasonably good at catching the fairly big fish, crustaceans, molluscs, echinoderms and so on, but they were not so good at collecting the tiny creatures, particularly those living actually in the muddy sediments on the bottom. This is mainly because they used nets with big meshes which allowed the bottom mud, and most small creatures, to pass through and escape.

The existence of the little beasts had, of course, been known about since the days of the *Challenger*, but they had never been taken in large numbers. This was rather surprising really, because it is very unusual to have an animal community not dominated by the smaller ones. The normal situation in any habitat is to have a sort of animal numerical pyramid, with lots of tiny beasts forming the base, rather fewer somewhat bigger creatures eating them and forming the next layer and so on until the apex of the pyramid is made up of very few large top predators. So where on earth were all the small beasts in the deep ocean?

In the 1960s a number of US marine biologists decided to find out – and the solution was not rocket science! They developed a sampling system that collected the bottom sediment – and everything that it contained – instead of throwing most of it away through holes in a net. The new gadget was called a box corer because it was half a tonne of stainless steel that was lowered to the deep sea floor to take a core of sediment in a box 50 cms square and about the same deep.

Considering what was happening in space research during the 1960s, culminating in Neil Armstrong landing on the moon in 1969, the development of the box corer seems pretty small beer. And I suppose it was. But to those of us working on inner space the development of the box corer was earth shaking. The first effect was that the samples it brought back suggested a bewildering diversity on the bottom of the deep ocean. This needed a bit of explaining because, for more than a hundred years, the deep sea floor had been considered an incredibly monotonous environment, not at all the situation that might be expected to produce lots of species. After leaning over backwards to try to find an explanation for this anomalous situation, scientists gradually realised that although the sea floor might be pretty uniform on scales of km (see Q17), at a much smaller scale, of centimetres to metres significant to the tiny beasts, the deep sea floor is extremely variable.

The present consensus is that a combination of disturbance by water currents, large beasts ploughing through the sediments and uneven distribution of food material in a generally food-poor environment, produce the ideal circumstances for the evolution of a huge diversity of tiny species. It has to be admitted that this conclusion is based on the analysis of a remarkably small number of box core samples, representing only something like 0.000000015% of the total area of the ocean floor! However, despite the huge error bars that have to be applied to these ideas, the best informed 'money' at the moment is on the deep sea floor being the most biodiverse environment on earth; even more biodiverse than the classic coral reefs or tropical rainforests. But because of the vast extent of the oceans, and the technical difficulties in studying it, the huge majority of the animals living on the deep sea floor have never been seen by human eyes – and never will be! A sobering thought.

32 What do animals in the deep ocean feed on?

Qs27 and 28 emphasized that, apart from the very special communities around hydrothermal vents and oil and gas seeps, all animal life in the oceans is dependent for its existence upon the plant growth in the thin sunlit surface layer. This means, of course, that all the herbivores in the oceans are also confined to the surface layers because in the deeper layers there are no plants for them to eat! As a corollary, all animals living beneath this layer have to be either carnivorous predators feeding on other animals, scavengers feeding on dead animal carcasses, detritivores living on small particles resulting from the breakdown of animals and plants – or various combinations of these. But the key to all this is *sinking*, because one way or another all these beasts rely on material sinking from the surface layers.

It is reasonably easy to imagine how fairly big things, like the bodies of fish, seals or even whales, can sink pretty quickly to the bottom. Even in the deepest parts of the ocean such large lumps will reach the sea floor in a matter of hours if they are not eaten by some hungry midwater animal on the way. But although these bonanzas certainly do reach the bottom from time to time, they represent a rather small proportion of the total input to the deep ocean. In contrast, the bulk of the material fuelling the deep sea originates in the bodies of the myriads of microscopic plant cells of the phytoplankton (see Q27), the slightly bigger bodies of the herbivorous zooplankton (see Qs26 and 27), and the tiny faecal pellets produced by these beasts.

Such tiny particles, with large surface areas relative to their mass or weight, would sink extremely slowly through the water column at rates measured in centimetres a day[17]. Even in moderate depths of a thousand metres or so, such slow-sinking bits and pieces would take weeks or months to reach the bottom – and most of them probably never get to the seabed because, like the large lumps, they would have to run the gauntlet of the hungry little beasts in midwater. But Q6 pointed out that the *average* depth of the oceans is almost 4,000 metres and the tiny sinking particles would take months or even years to reach such depths.

---

17 If you find this concept of sinking rate being affected by the surface area to mass ratio difficult to grasp, think of what would happen if you threw a big lump of cheese, say the size of a brick, into a pond or the sea. It would sink like the proverbial stone, and almost as fast as if it were actually a brick, slowed down only very slightly by the friction between the water and its surfaces. Now imagine what would happen if you first cut the cheese up into hundreds of tiny pieces – or even grated it – and then threw the individual bits into the water. The total surface area of all the bits is now huge compared with that of the undivided brick-sized lump, so the frictional resistance to their passage through the water is also very much bigger. Consequently, the little bits sink more slowly than the big bits.

Until only about 30 years ago, deep sea biologists thought that this was exactly what did happen and that this would have two clear results. The first was that with increasing depth there would be less and less food available so that animal life in general would become more and more sparse. The second was that any seasonality in the surface waters, with strong summer-winter contrasts in the amount of plant growth at mid-latitudes, would have long since disappeared by the time the tiny particles reached the bottom. In this view, the deep ocean floor was subjected to a constant but very thin 'drizzle' of minute food particles with only minor differences from place to place and from time to time. In summary, the deep sea floor was seen as a very food-poor and extremely monotonous place. This seemed to fit pretty well with what had been discovered about life on the deep ocean floor over the previous century or so; as you sampled deeper and deeper in the ocean you found fewer and fewer animals belonging to fewer and fewer species.

But there were a couple of rather inconvenient facts that seemed to run contrary to this nice simple idea. For one thing, in the 1960s American scientists discovered that there were many more different beasts among the little ones living in the sediment than anyone previously had dreamt of (see Q31). This new-found biodiversity ran totally counter to the idea of a very monotonous deep sea floor and, for several years, biologists metaphorically jumped through hoops to try to explain it using what seem to be more and more far-fetched ideas. Then there was another baffling stumbling block; seasonal breeding among deep sea animals!

Sampling the deep ocean frequently enough to study details of the life history of individual species was, and still is, very difficult. Consequently, knowledge of the reproduction of animals living close to the bottom of the deep ocean has always been rather sparse. Nevertheless, a few intrepid scientists persevered in this sort of study and painstakingly worked out the reproductive biology of a small number of deep living species. As expected, most of them reproduced fairly randomly, with no particular peaks or troughs in the numbers of animals in a population reproducing at any one time. But amazingly, a few didn't! Instead, a small number of animals, including some small echinoderms and molluscs, seemed to show defi-nite seasonality in their reproduction, producing eggs and sperm more or less in line with what was going on thousands of metres above them. How on earth could this be? After all, everyone knew that the deep ocean is permanently and absolutely dark, so there could be no 'clue' from seasonal changes in day length or light intensity. Similarly, the temperature and salinity of the near bottom water was known to be virtually constant, probably over millennia let alone months! So what could possibly be the link between the surface layers and the bottom? The answer turned out to be what is now known as 'marine snow'.

As the name suggests, marine snow consists of easily visible particles, or 'flakes', sinking through the water column. They were first noticed by scientists diving in deep submersibles in the Pacific in the 1950s, but have since been seen, photographed and even collected in many parts of the ocean. They are made up of many different materials, including a sort of 'snotty' mucus produced by filter-feeding creatures to collect tiny food particles in midwater, the dead remains of tiny animals, some caught up in the mucus, various bits of poo in the form of faecal pellets and, perhaps above all, the remains of surface-dwelling phytoplankton, or phytodetritus. But a common feature is that the 'flakes' are large, sometimes centimetres across, and much bigger than most of the individual bits and pieces that are included. This means, of course, that they sink much more rapidly, sometimes hundreds of metres in a single day.

The mechanisms leading to the production of marine snow are by no means all understood, but one factor seems to be the deterioration and eventual death of plant cells, particularly diatoms (see Q27).

As a result, during the spring phytoplankton bloom at mid-latitudes, there is a massive deposition of detritus, formed mainly of diatom remains, which sinks rapidly through the water column picking up more material as it does so. It reaches the sea floor as a food-rich gunge or fluff, which is easily visible in sea floor photographs. This injection of maritime compost clearly provides a strong seasonal signal to the deeper layers, presumably explaining the otherwise puzzling seasonal breeding. But it may also help to explain the unexpected diversity of the deep sea floor. By being wafted backwards and forwards by near-bottom water currents and being moved around by the animals crawling across and through the sediments, the phytodetritus becomes very unevenly distributed, settling around mounds and in small depressions and grooves in the seabed while leaving other areas completely bare. So at the scale of the tiny animals that make up the bulk of the species-rich bottom-living communities, the deep seafloor is by no means the boring, monotonous environment that scientists thought it was. Instead, it is an ever-changing mosaic of food-rich and food-poor patches, from time to time disturbed by large fish and invertebrates or major food falls, and providing just the conditions to encourage the evolution of many different species over the eons of the oceans' existence.

33 What are jellyfish?

Jellyfish are animals belonging to the same group, the Cnidaria, that contains the sea anemones, sea pens, sea fans and corals. A unifying characteristic of the Cnidaria is that they possess stinging cells that can inject a toxin that is often painful and is sometimes deadly.

Most cnidarians live a stationary life as adults, though they may spend days or weeks as juveniles drifting around in the water column. In contrast, jellyfish spend most of their lives in midwater, though at least some of them have a quite distinct phase of their life cycle when they are attached to rocks, seaweeds or other animals. At this stage, called a scyphistoma (from two Greek words meaning cup-mouth), they are unrecognisable as jellyfish, but look a bit like a tiny ice-cream cone just a few millimetres across. The cone is made up of a series of little discs piled on top of one another, which are budded off one at a time to become potential new jellyfish. Lots of them are eaten by fish and other creatures, of course, but if they survive they grow into the familiar jellyfish we see in the surface waters, slowly propelling themselves along by pulsing their bell-shaped bodies like opening and closing umbrellas. As they are doing so, they are feeding on small planktonic organisms including crustaceans, squid and even fish. These are caught using the stinging cells on tentacles hanging from the outer edge of the bell or on fleshy lips surrounding the mouth in the middle of the concave side of the bell.

Whereas the scyphistoma stage reproduces by budding, the jellyfish, or medusa, stage reproduces sexually, with most species having the sexes separated so that each jellyfish is either a male or a female. The eggs and sperm are shed into the sea and produce tiny larvae that hopefully settle on a suitable substrate to form the next generation of scyphistomas and jellyfish.

There are some hundreds of different jellyfish species in the oceans, but the most frequently seen ones, particularly in north Atlantic waters, are the so-called common or moon jelly and the lion's mane jelly. The common jellyfish, *Aurelia aurita*, reaches a diameter of 30-40 cm and is usually a translucent blue-ish or green-ish colour. It has a fairly flat bell with lots of short tentacles all round the outer margin and the lips of the mouth are drawn out into four prominent lobes. But *Aurelia* is always instantly recognisable because of the four prominent purple or lilac-coloured moon-shaped gonads arranged in a square pattern in the middle of the umbrella, providing it with the alternative name of moon jelly.

Aurelia stinging cells produce a fairly weak toxin that can nevertheless be very irritating if you get it in your eye or on a particularly sensitive part of your skin. But it is nothing in comparison with the sting of the other common northern jellyfish. The lion's mane, *Cyanea*, is much bigger than *Aurelia*, reaching a bell diameter of almost 2 metres in the Arctic, though a bell half this size would be quite big in European or North American waters. It occurs in two distinct colour varieties, possibly distinct species, one blue, the other predominantly red or orange. But apart from the colour, it is immediately distinguishable from *Aurelia*, or any other jellyfish you are likely to see, because of the dozens of long, fine tentacles that stream out behind it as it swims slowly through the water. These can extend to 4 or 5 times the diameter of the bell, so in murky waters a fish, or a swimmer, can come into contact with the tentacles before they see the jellyfish. But they would soon know, because the rows of stinging cells mounted along the tentacles inject a very powerful poison that can be extremely painful.

Jellyfish can occur in enormous swarms and can sometimes clog fishing nets. Fishermen are, understandably, fairly anti-jellyfish, not only because they interfere with their catches, but also because the stinging cells can survive drying out. Consequently, nets that have been exposed to jellyfish swarms but have been out of use for some time can sting fishermen quite badly when they are re-wetted, even if there are no jellies about!

34 What is the Portuguese Man-o'-war?

The Portuguese Man-o'-war, *Physalia physalis,* also known as the blue-bubble or bluebottle, is a jellyfish-like creature that lives at the very surface of the ocean supported by a blue-ish translucent gas-filled float up to 20 or 30 cm long. Although it is included in the same major group (the Cnidaria) that contains the jellyfish, the Portuguese Man-o'-war belongs to a quite separate division, the Siphonophora, with very different characteristics. Whereas true jellyfish are individual animals that fulfil all the basic functions of feeding, moving and reproducing, the siphonophores, including *Physalia,* are colonial systems containing several types of specialised individuals that could not survive in isolation. So beneath the float of a Portuguese Man-o'-war hang a number of individual 'polyps', specialised either for reproduction or for feeding.

The feeding polyps have contractile tentacles that may reach 10 metres in length, though they are usually much shorter. They carry hundreds of stinging cells that can inject a very powerful toxin used to immobilise the normal prey of crustaceans and small fish. The toxin causes excruciating pain in humans and has been known to result in death, though probably only in individuals with pre-existing heart or respiratory conditions. The Portuguese Man-o'-war is primarily a warm water creature and sometimes occurs in such large numbers that its floats are cast ashore in hundreds looking like blue bubbles, hence one of its names. But since it is carried along by winds and currents, it is a regular, but occasional, visitor to British shores where its presence usually causes unjustified panic in the media.

The Portuguese Man-o'-war, like the true jellyfish, is eaten by turtles which are apparently totally immune to the toxin.

35 Does anything eat jellyfish?

Yes, but not much!

Jellyfish, like their namesake, jelly, consist mainly of water with rather little nutritional value. Consequently, anything making a living out of eating jellyfish couldn't afford to expend much energy finding them. So it is not too surprising to find that jellyfish eaters *par excellence* are turtles and the ocean sunfish (see Fig 11), neither of which do very much other than drift around in the surface waters of the ocean, hopefully, from time to time, running into the odd jellyfish! Although the stings of many jellyfish are toxic, some extremely so (see Q34), sea turtles and sunfish seem to be totally immune.

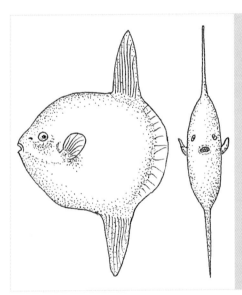

Figure 11
The Ocean sunfish, *Mola mola*, side on and head on. The sunfish's body can reach 8 feet (2.5 m) across and weigh more than a tonne. The tail fin seems ridiculously small for a beast of this size, while the upper (dorsal) and lower (anal) fins are enormous. As the fish swims lazily along, its dorsal fin may cut through the surface and look deceptively like a shark.

Interestingly, the only insect to have successfully colonised the open ocean environment, the sea skater, *Halobates* (see Q36), also feeds on surface-living jellyfish like the Portuguese Man-o'-war (Q34).

But the other major eater of jellyfish is – mankind! Jellyfish have been an important part of Chinese cuisine for many hundreds of years and became the object of an important fishery in south-eastern Asia when the Japanese became interested in them in the 1970s. Currently, some 300,000 tonnes wet weight of jellyfish are marketed each year. These are principally species belonging to one family of jellies, the Rhizostomeae, which are relatively firm bodied.

36 Are there any marine insects?

Yes! Five species of the genus *Halobates*, called sea striders or ocean skaters, spend their entire lives on the surface of the open ocean. They are close relatives of the familiar pond skaters of fresh water and are widely distributed in the subtropical and tropical areas of the world ocean between about 40°N and 40°S. The genus contains about another 40 species found in sheltered inshore waters of the tropical Indo-Pacific.

Since, like their freshwater cousins, the ocean skaters are confined to the very surface film, they have to make do with this environment for all their biological needs. Their eggs are laid on almost anything that floats, such as seaweed, wood or bits and pieces thrown overboard from ships. They feed on floating jellyfish like the Portuguese Man-o'-war and possibly also on floating fish eggs and small crustaceans trapped in the surface film. In turn, they are eaten by fish and probably form important elements in the diet of surface feeding seabirds.

An interesting but unanswered question about *Halobates* is why its five oceanic species are the only totally marine representatives of the almost 1 million insect species that have colonised all other environments on earth? No one knows, but it is possibly because a related jointed-legged group, the crustaceans (see Q43), got there first!

37 What is bioluminescence?

Bioluminescence, sometimes referred to less correctly as phosphorescence, is light produced by living organisms. A few land animals are capable of producing bioluminescence, notably fireflies and glow-worms, but it is much more common among marine organisms. Members of lots of animal groups in the sea produce bioluminescence, including quite a few in the surface layers. But the phenomenon is particularly common in the sunlight-deprived zones of the deep ocean. For example, more than 70% of the midwater fish and shrimp species living deeper than 500 metres in the North Atlantic are luminous, indicating the importance of the phenomenon in this environment.

Although the precise method of producing light varies in different groups, the basic mechanisms are common to all of them. It involves the oxidation (similar to burning) of a fuel, in this case an organic molecule referred to as luciferin, in the presence of an enzyme called luciferase. The oxidised luciferin then decays back to its original form, releasing energy in the form of cold light rather than as heat, which is more usual in other oxidation processes like burning coal or wood.

The light produced in this process is usually blue or greenish, and is generally not very bright, though it must be highly visible in the pitch black ocean depths. Its functions are not always obvious, but they include camouflage, scaring potential predators, finding a mate and even attracting food!

Shellfish

38 What are shellfish?

'Shellfish' is a general, non-scientific term that includes all the animals in two quite distinct groups; the crustaceans (Q43) and the molluscs (Q40). It is used particularly for the edible members of the two groups and refers to the fact that most, though not all, have inedible shells.

39 Why does eating shellfish make some people very ill?

There are two possibilities: first food poisoning, second shellfish allergy.

Food poisoning, as the name suggests, results from eating something that contains toxins. These are usually the result of bacterial action either because the food has not been treated hygienically and has therefore picked up pathogens of some description, or because the food material has started to decompose. This could, of course, be the cause of food poisoning from eating shellfish, but a further possibility is that the shellfish itself could have accumulated toxins naturally.

As Q38 pointed out, the term shellfish covers two major animal groups, the molluscs and the crustaceans; the problem of accumulating toxins naturally is associated with molluscs and particularly with those bivalve species (such as mussels, oysters, cockles and scallops) that feed by filtering food particles out of the water. The food particles are primarily tiny phytoplankton cells such as diatoms, dinoflagellates and cyanobacteria. Several species produce toxins, but in small numbers do not significantly affect the molluscs. However, they may be involved in harmful algal blooms (often called red tides) and can be so abundant that their toxins build up in the bodies of the molluscs exposed to the bloom to reach dangerous levels if eaten by humans. The effects can be relatively mild but can also involve nausea, vomiting, diarrhoea and even paralysis, in extreme cases even resulting in death.

Anyone can suffer from food poisoning, but a shellfish allergy is quite different. Shellfish allergy is a true food allergy, that is a reaction of the body's immune system to some feature of the food, usually a specific protein. In western societies about 1 in 70 people have a true food allergy, with problem foods including peanuts, wheat, soya, eggs, milk, fish – and shellfish. The effects of shellfish allergy can be extremely dramatic and serious but are usually restricted to unpleasant, but not necessarily life-threatening, symptoms like coughing, sneezing, vomiting and urticaria.

Although the two animal groups making up shellfish are not closely related, many people who react badly to eating crustaceans such as shrimps, crab and lobsters, may also find that they have a similar reaction to eating molluscs, and vice versa. Once a shellfish allergy has been identified it is therefore wise for the sufferer to avoid eating all shellfish. Since shellfish are not normally well represented in the diets of young children it is also as well for parents to keep a careful watch on youngsters when they are first exposed to food containing molluscs, crustaceans – or even insects!

40 What are molluscs?

Molluscs make up one of the major divisions (the phylum Mollusca) into which biologists divide the animals lacking backbones. Along with the crustaceans (see Q43), we group the molluscs under the unscientific but useful term shellfish (see Q38).

The Mollusca contains about 200,000 separate living species grouped into eight different classes, but three of these are restricted to the deep sea. Of the remaining five classes, three contain the vast majority of molluscs that most of us are interested in, either because we eat them or because they are a nuisance.

The biggest class by far is the Gastropoda, containing about 150,000 species found in the oceans, in fresh water and on land. They include all the snails, slugs and limpets. Most marine representatives crawl around on seaweed, rocks or the sea floor and feed by scraping organic material from the surface they are moving over. Many of them have heavy shells, but quite a few have reduced their shells or even lost them altogether. A small number of species, often grouped together under the general term pteropods (meaning wing-footed), have even taken to a midwater lifestyle. These specialist swimmers include the sea butterflies or flapping snails, so-called because the foot of their more typical relatives has become modified into a pair of wing-like lobes with which the tiny animals propel themselves through the water. After death, their thin calcareous shells, each several millimetres across, sink to the sea floor and in some places can dominate the sediments as a pteropod ooze, often referred to by deep sea biologists as 'rice crispies'.

The second biggest class is the Bivalvia, which, as the name suggests have a shell in two parts or valves, hinged together. The bivalves include all the clams, oysters, scallops and mussels and are found in the sea and in fresh water, but not on land. Unlike the gastropods, most bivalves feed by filtering small particles out of a current of water wafted through the space between the shells.

The third class, the Cephalopoda, contains less than a thousand species, but includes all the squid, cuttlefish and octopuses. All are marine. They are by far the most active of all the molluscs and most are predators on fish, crustaceans and even other molluscs. They have all reduced their shells: to an internal plate (the familiar cuttlebone beloved of cage bird enthusiasts) in the cuttlefish, to a narrow cartilaginous quill in the squids, and to nothing at all in the octopuses – explaining how they are able to insinuate their bodies into unbelievably restricted spaces!

The remaining two molluscan groups are not very familiar to non-zoologists. One, the Scaphopoda or tusk shells – so named because their shells, up to a few cm long, vaguely resemble elephants' tusks – live on soft sediments generally below the tide marks and feed on microscopic prey. The other group, the polyplacophorans or chitons, are slightly more familiar because several species live inter-tidally. They are rather primitive molluscs with the shell made up of eight separate overlapping and articulated plates, allowing the animal to flex when crawling over uneven surfaces and even to roll slowly into a ball if dislodged – a bit like a woodlouse!

41 How big are giant squid?

Squid are cephalopod molluscs, related to the snails, slugs and oysters (see Q40).

The existence of giant members of the group has been known about for many years, possibly centuries, though the first proper scientific description was published as recently as 1857. Almost inevitably, giant squid gave rise to amazing stories told by returning sailors of huge and frightening monsters capable of grasping whole ships in their tentacles and dragging the vessels, and their crews, to a nasty death in the depths of the ocean. These stories were invariably total nonsense, but they were based on very real creatures.

There are several species in the oceans deserving the name giant squid, all placed by biologists in the genus *Architeuthis*, meaning 'ancient squid'. However, the main claimant to the title seems to be *Architeuthis dux*, meaning 'the chief ancient squid'. This squid is recorded from all the world's oceans, including the

North Atlantic, off Newfoundland, Norway and the northern British Isles, from around the Azores and Madeira, off Southern Africa, from the north Pacific around Japan and from the south-western Pacific around New Zealand and Australia. So no matter where you are in the oceans you can be pretty sure that there are giant squids not far away.

But despite their widespread distribution, rather little was known about the biology of giant squid until fairly recently because they seem to live mainly at depths of 500 to 1,000 metres in the oceans and are difficult to collect. Consequently, most specimens examined by scientists have been dead and rotting animals floating at the surface or cast ashore by wind and waves. Other bits and pieces, particularly their jaws which are made of fairly indigestible horn-like material, have been obtained from the stomachs of sperm whales. Some of these whales carried scars of squid arm suckers on their skins, often interpreted as evidence of titanic struggles between these two leviathons of the oceans. Based on this material, estimates of the maximum size attained by the squid have reached 20 metres, about 65 feet, of which 4-6 metres would have been the main body of the squid or mantle.

The situation has improved somewhat in recent years with the capture or filming of living specimens particularly around New Zealand and Japan, though the giant squid is still one of the most mysterious of deep sea animals. The new information suggests that the early estimates were probably exaggerated and were based on measurements of body parts, particularly the tentacles, which had been unnaturally stretched. Modern estimates suggest a maximum mantle length of about 2.25 metres (7.5 feet), with an absolute maximum length, including the tentacles, of 13 metres (43 feet) for females and 10 metres (33 feet) for males. Not only are the biggest females considerably longer than the males, but giant squid ladies also apparently reach a much greater maximum weight at an estimated 275 kg (606 pounds) compared with only 150 kg (330 pounds) for the males.

So do these creatures deserve their reputation as terrifying monsters capable of destroying entire ships and possibly sperm whales? Absolutely not! All the evidence seems to suggest that giant squid, or at least *Architeuthis dux,* is actually a deep ocean pussycat, certainly huge, but something of a couch potato rather than a go-getting hunter. Living, as it does, at depths in the ocean where food is not very abundant, it would make little sense for such a vast animal to rush around looking for something to eat and using up lots of energy. Instead, it seems much more likely that giant squid spend most of their time simply drifting in the water column and hoping that some suitable food item, particularly a fish or another squid, would drift within reach of their not-very-muscular arms so that they can grab it and eat it. So what about the stories of life and death struggles between sperm whales and giant squid? Sorry, but that is also nonsense. It seems that no self-respecting sperm whale would come across a giant squid during its deep dives and look at it as anything other than a very tasty and totally harmless morsel to eat. A giant squid finding itself seized in the jaws of a sperm whale might well grab hold of its attacker with its arms and leave impressive scars on its skin with its suckers. But its efforts would be a lost cause, and no matter what the squid did its fate would be to end up in the stomach of the whale! Essentially, a squid, no matter how big it is, is no match for a sperm whale.

But there is a slight caveat to this rather dismissive story of giant squid. It is just possible that there is a beast in the deep ocean that does justify the giant squid mythology. So if you think giant squids are big, have a look at the Colossal Squid, Q42.

42 What is the colossal squid?

The colossal squid, *Mesonychoteuthis hamiltoni*, is believed to be the heaviest squid on the planet and also the earth's largest invertebrate, that is an animal without a backbone.

This species has been known since 1925 when it was described from two bits of tentacle found in the stomach of a sperm whale. But no one had any idea how big the species could be until the last few years when a number of successively bigger and more impressive specimens were collected in Antarctic waters. Finally, in February 2007, fishermen using long lines in the Ross Sea brought aboard the largest specimen ever obtained when it attacked an Antarctic toothfish already caught on one of the hooks, and refused to let go. It weighed in at 495 kg (1,091 pounds) and was estimated at about 10 metres (33 feet) long, though, after it had been frozen and then thawed, scientists at New Zealand's national museum found that the tentacles had shrunk a good deal and it then measured a relatively modest 4.2 metres (14 feet).

Measuring these beasts accurately is very difficult, partly because they are pretty floppy anyway, but also because by the time they get into the hands of scientists used to dealing with these things they are often not in the best condition. So although there have been statements in the popular press suggesting that colossal squid may equal or even exceed the length of giant squid (*Architeuthis dux*, see Q41) this does not seem to have been corroborated. The colossal squid is certainly a much more stocky, and therefore heavier animal, but it probably comes in second in the length stakes!

The colossal squid, like the giant squid, lives in the deep sea, in this case possibly down to depths of 2,000 metres (6,000 feet) or more, where it feeds on fish, crustaceans and even other squid. Because it is much more muscular than

Architeuthis, the colossal squid may be a more active and aggressive predator. Nevertheless, neither species is likely to pose a serious threat to whales and the colossal squid, like the giant squid, is certainly eaten by sperm whales and some deep-sea sharks.

Whereas the giant squid is very widely distributed in the world oceans, colossal squid seem to be confined to southern hemisphere waters, particularly south of 40°S. However, so few specimens have so far been collected that our present knowledge of this species, including its distribution, may turn out to be quite wrong.

43 What are crustaceans?

The crustaceans are all those animals without backbones that we group together, along with the molluscs (see Q38), as shellfish.

Crustaceans form just one part of a much larger animal group, the phylum Arthropoda (meaning 'jointed legged'), which also includes the arachnids (like ticks, mites, spiders and scorpions), all the millipedes and centipedes, and, finally, the biggest group of all, the insects. All these creatures have their supporting skeleton on the outside, enclosing all the soft bits. And because this outer skin can't expand continuously, and in many species is absolutely hard and rigid, all arthropods have to shed or moult their skin periodically in order to grow.

There are almost 1 million described species of insects, that is more than half of all known animal species put together (but see Q31). Compared with this, the estimated 52,000 crustacean species are pretty small beer, but in the seas, crustaceans are extremely important.

Apart from a few very specialised insect species inhabiting the ocean's surface film (see Q36), insects are entirely lacking from the marine environment. In contrast, although there are quite a few freshwater crustaceans (like crayfish) and rather fewer terrestrial ones (like the familiar woodlice or pill bugs), the crustaceans are mainly marine, with representatives in all oceans and at all levels from the surface to the bottom of the deepest trenches. In fact, they can be considered the insects of the seas. But living in the seas instead of in air has allowed the crustaceans to avoid some of the constraints, particularly on size, that limit the range of insect form. So crustaceans can be much bigger than insects and the largest known crustacean, the giant Japanese spider crab, has a leg span of up to 4 metres. An insect that size would be truly awesome!

The crustaceans that we tend to be most familiar with are those we eat, like lobsters, shrimps, prawns and crabs (see Qs 44, 45 and 46), though most people are aware that the group also includes things like krill, the food of the big whales (see Q47). However, interesting and important though these groups are, the crustaceans also include lots of less familiar groups of small and not very obvious creatures, some of which fulfil particularly crucial roles in the economy of the seas. Principal among these are the copepods, a name that means 'oar-footed', alluding to the fact that many species seem to row themselves through the water. The copepods are tiny shrimp-like creatures, usually with a cigar shaped body with long feelers or antennae at one end and a short stubby little tail at the other. They range in size from about a millimetre to more than a cm long, but a fairly typical one will have a body about the size and shape of a grain of rice.

There are more than 20,000 described species of copepods and while many of them live entirely in midwater, even more live on or in the sediments and some are specialised parasites, particularly on fish, mammals and other crustaceans.

It is the midwater ones that are particularly important. Most species living in the surface layers are specialised feeders on phytoplankton (see Q27), many of them filtering out individual microscopic plant cells as they propel themselves through the water. In turn, they provide food for other invertebrates and fish; for example, one species, *Calanus finmarchicus*, is a major food of herring and therefore its welfare is crucial to any healthy herring fishery.

However, some copepods occur in such dense swarms that they are eaten by some of the whalebone whales so, just like the herbivorous krill being eaten by blue whales, copepods provide an extremely efficient link between the tiniest plants and some of the biggest animals in the oceans.

44 What is a lobster?

On the face of it, this seems a bit of a daft question. Surely, everybody knows a lobster when they see one, don't they? Well, they might; but as with so many names applied to edible animals, there is room for a lot of misunderstanding and the general name 'lobster' is applied to well over 70 different species in different parts of the world. So the answer to the question depends quite a lot on where you happen to be at the time. Let's try to unravel the confusion a bit.

First, all lobsters are crustaceans, belonging to the same big section of the group that also contains all the crabs, shrimps and prawns. All of them have five pairs of legs and are given the general name Decapoda, which simply means 'ten legged'. The legs are used for walking, of course, but some of them are also used for feeding and are provided with pincers or claws to gather food and pass it to the mouth. And whether these claws are big or small provides the main distinction between the two main types of lobster, basically the clawed lobsters – and the rest.

In Europe and North America we are familiar with seeing the typical blue-black lobster with two huge front claws on the fishmongers slab or, more often these days, on the fish counter of our supermarket[18]. The clawed lobster is known in France as 'homard' and in Germany, Denmark, Norway and Sweden as 'hummer', from which it gets its scientific genus name *Homarus*[19]. There are two species, one on either side of the Atlantic and both living in relatively shallow water, that is, generally on the Continental Shelf (see Q13). The eastern or European one, *Homarus gammarus,* is found from northern Norway in the north to Agadir on the coast of Morocco in the south. It is also found as far west as the Azores and eastwards into the Mediterranean and a bit of the Black Sea. It is fished fairly intensively throughout this range, mainly with baited traps (lobster pots or kreels), and is sold exclusively in European and North African markets. The western species, *Homarus americanus,* occurs from Labrador south to Cape Hatteras and is known by various local names, but particularly as the Canadian, Maine, northern or American lobster. Like its European cousin it is caught in traps, but it is also frequently trawled. It is taken in huge numbers and many north American lobsters are exported to European markets. The two species are indistinguishable other than by an expert, though the North American form tends to grow rather bigger.

The two big clawed lobster species have a number of rather smaller close relatives which also have well-developed claws on the ends of the first pair of legs. Several of these are fished in various parts of the world, but by far the most commercially important is the Norway Lobster or Dublin Bay Prawn, known much more generally in Britain by the name scampi (see Q45). None of them are likely to be confused with their big brothers, but there are plenty of other species that could be.

By far the biggest group are the spiny lobsters, so-called because they tend to be far more spiny than the clawed lobsters, particularly on the carapace (the big shell above the legs) and on the tail segments (see Fig 12). They belong to the family Palinuridae and the genera *Palinurus, Panulirus* and *Jassus.* They occur in the temperate and tropical waters of all the world's oceans and are landed in fish markets all over the world. If you can see the whole animal, spiny lobsters are very easy to distinguish from the clawed lobsters, mostly because they totally lack big claws. But they also tend to have much thicker and spinier antennae (feelers) and are much more brightly coloured, often including blues, greens and yellows.

---

18   Clawed lobsters turn red only after they are boiled.
19   The English name lobster is apparently derived from an old English word loppestre or lopystre, which seems to have been applied to more or less any crustacean.

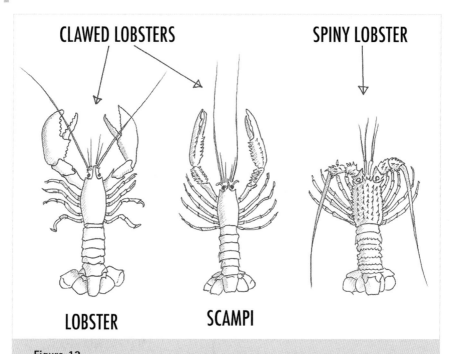

Figure 12
The two basic types of lobster, the clawed lobsters on the left and spiny lobster on the right. The true lobster and scampi are shown here at the same size, but scampi grows to less than half the maximum size reached by lobsters.

The third group of potential 'lobsters' belong to a family called the Galatheidae. Scientists refer to galatheids as 'squat lobsters' because, instead of holding their tails straight out behind the main body like any self-respecting lobster, or even shrimp, the galatheids squat, that is, they fold the tail under the body, almost as if to hide it. Squat lobsters are found throughout the shallow parts of the world oceans, including around European and North American coasts, but they are exploited commercially only in South America and particularly in Chile. The name 'squat lobster' is unlikely to encourage many buyers so the galatheids are marketed under a variety of other names, particularly 'langostinos', the name they are known by in Chile.

All these 'lobsters' are perfectly acceptable as human food, though my personal view is that most of them are overpriced. Given a choice, however, I prefer the taste of the relatively slow-growing, cold water, clawed lobsters (*Homarus* and *Nephrops*) to the warmer water spiny and squat lobsters. Others might disagree with me and prefer the taste of spiny lobsters or even squat lobsters. Either way, you might think, like me, that it is quite nice to know what you are buying or eating. I've already told you how to distinguish the different groups if you can see the whole body; claws or no claws, lots of spines or few spines, straight tail or turned-under tail. But lobsters are often sold as tails without the front ends attached to them. So how can you tell them apart without the head end? The answer: look at the other end, the tip of the tail (see Fig 13).

The end of the tail of all lobsters (and, for that matter, all decapods) consists of five separate plates forming a sort of fan, with a central plate (called a 'telson' by crustacean scientific nerds, like me), flanked on either side by paired 'uropods'. The whole thing looks a bit like a butterfly with a central body and paired wings on either side, but it can tell you exactly which group its owner belongs to. Here's how.

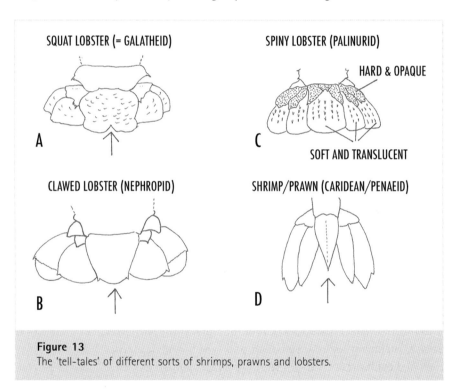

Figure 13
The 'tell-tales' of different sorts of shrimps, prawns and lobsters.

If the central plate, the telson, has a V-shaped notch in the middle of its rear edge (Fig 13A), it is a squat lobster. If the telson has an uninterrupted rear edge and all the branches of the tail fan are hard and opaque throughout their length (Fig 13B), you are looking at a clawed lobster, either a 'proper' lobster, one of the *Homarus* species, or one of the smaller species like scampi. Finally, if the flattened branches of the tail fan are a bit bendy, or leathery and pliable, translucent in the hind part and opaque in the forepart, often with quite a few little spines (Fig 13C), it is a palinurid or spiny lobster.

But if the central plate, the telson, is sharply pointed (see Fig 13D) rather than rounded, you are looking at a shrimp or a prawn and emphatically not at a lobster, a spiny lobster or a true scampi. Since 'lobsters' and 'scampi' command a premium price, if you are offered a pointed-tailed crustacean as either of these, challenge the fishmonger or restaurateur, and see what they say. I do! Good luck.

45 What is scampi?

The word scampi is simply the plural of the Italian word scampo, meaning shrimp. So scampi should mean shrimps. But the beast that we all refer to as scampi is emphatically not a shrimp. Nor is it a prawn, despite the fact that it is called a prawn in Scotland and one of its English names is Dublin Bay Prawn. It is actually a lobster (see Fig 12), and its other English name is Norway Lobster, reflecting its Latin name *Nephrops norvegicus*. It was first properly described from Norwegian waters in the eighteenth century by the famous Swedish biologist Carl Linnaeus, but it is now known to occur throughout the north-eastern Atlantic from Norway in the north to Morocco in the south and also in the Mediterranean – hence its Italian epithet.

The Norway lobster lives on muddy bottoms where it spends most of its time in burrows, coming out only to eat or to interact with its fellow Norway lobsters, mainly to mate. Although it can be found in fairly shallow water, sometimes in no more than 30 metres, it generally prefers rather deeper water and can be found as deep as 700 metres or so. Growing to a maximum length of about 40 cm, including the claws, it is quite a bit smaller than its 'proper' lobster relative (see Q44). Also unlike proper lobsters, which are blue-black in life and only become red when they are boiled, Norway lobsters are an attractive orange colour when alive.

Scampi became popular in the United Kingdom in the late 1950s and 1960s when holiday makers came across it in the Mediterranean. Sadly, many people still think of it as a rather exotic import from warmer climes. This could not be further from the truth. By far the best and biggest scampi in the world are fished from British, and particularly Scottish, waters. If you enjoy a plate of scampi in Spain, France or Italy, the chances are that it was caught in the cold, clear waters off the west coast of Ireland, Scotland or Norway[20].

---

20 I am delighted to have to qualify this rather jingoistic statement a bit as a result of a visit to the fish market in the Spanish Moroccan port of Ceuta in February 2009. It is a fantastic market with a range and quality of seafood to warm the cockles of any old oceanographer's heart, with a remarkable variety of fish species along with squid, mussels, crabs, lobsters and even barnacles. But the crowning glory for me was the scampi. I have never seen bigger, or fresher, scampi anywhere in the world – including Scotland. I can't offer a better accolade.

But here is one other note of caution when ordering scampi in foreign countries. In France our scampi is their *langoustine*[21]. The similar sounding *langostino* in Spanish refers to a very specific creature, the camarote prawn, in Spain, but to squat lobsters (see Q44) in South America and the United States. The Spanish call scampi *cigalas*, while Americans travelling in Europe are totally thrown by scampi dishes because at home the term scampi is applied to more or less any shrimp-like beast cooked in garlic and butter. Confusing, isn't it?

46 What is the difference between a shrimp and a prawn?

Not a lot, but it is almost impossible to define the two terms precisely because they are used in different senses in different parts of the world. The first thing to say is that shrimps and prawns all belong to the same ten-legged (decapod) group of crustaceans that also contains all the crabs, lobsters and hermit crabs, but shrimps and prawns have much smaller claws and spend quite a lot of time swimming whereas most of the others prefer to crawl around on the bottom.

The Brits are a bit unusual in making a distinction between shrimps and prawns[22]. For example, the French call everything vaguely shrimp-like 'crevettes', with adjectives like 'grise' (grey) or 'rose' (pink) to distinguish between the different species. In Spain the general term for both shrimps and prawns is 'camaron' ('camarão' in Portuguese), though the Spanish also use the term 'gamba' for some species.

So what do the terms shrimp and prawn mean in Britain? Well, basically 'shrimps' are small and 'prawns' are a bit bigger. Unfortunately, nobody can tell you exactly what 'big' and 'small' mean in this context, so we have to dig a bit deeper for an answer. Without using the Latin names (why should they?), the British seafood trade has traditionally used the term shrimp to refer to decapods belonging to the family *Crangonidae*. Sorry about that, but it doesn't have a common name. It contains a number of different species, all coloured a sort of dirty brown or grey in life, though, like most crustaceans, they tend to turn red when boiled. They are all small, no

---

21   The French name *langoustine* means 'little langouste'. But this is almost as inappropriate as the name scampi, because the *langouste* is the spiny lobster, and belongs to a quite different family from that containing the true lobster, including scampi (see Q44). Just to make things a little more complicated, Norway lobsters are also known in various areas as deep sea lobsters, Icelandic baby lobsters, deep sea tails and even deep sea dainties (see *Lobsters of the World - an illustrated guide*, Austin B. Williams, Osprey Books, 1988).

22   In Scotland the term prawn is used quite differently (see Q45).

more than 7-8 cm (3 inches) long and usually a good deal less, and are slightly flattened from top to bottom rather than from side to side. They live on fairly soft parts of the seabed, particularly on sand, and the most abundant one, *Crangon crangon,* is usually called the brown, grey or sand shrimp. It is found all round the British (and most north-western European coasts) in waters down to about 20 metres deep, but is perhaps best known from Morecambe Bay. In contrast, the term 'prawn' is usually applied to species belonging to another family without a common name (other than prawns), the Palaemonidae, and particularly to *Palaemon serratus,* the so-called Common Prawn, though in recent years it has been anything but common. The common prawn can be considerably bigger (up to 4-5 inches, 10-11 cm long) than the brown shrimp and is much more attractively coloured. In life they are largely almost glass clear, with patches of red, blue and yellow, but boiling turns them the usual dead crustacean red colour. Prawns also have a much more prominent beak or rostrum, that is the spine sticking forward in front of the eyes.

Unlike the brown shrimp, they are not confined to sandy sea floors but are also found around rocks and seaweed, being the ones usually caught by children fishing in rock pools between the tide marks. *Palaemon,* like *Crangon,* lives in fairly shallow water, but some deeper living species are also found in European markets. These belong to another family without a common name, the *Pandalidae,* far more similar to the *Palaemonidae* than to the *Crangonidae* and, accordingly, usually called prawns, often distinguished as deep sea prawns, northern prawns or deepwater prawns. Just to add to the confusion, however, they are also sometimes called pink shrimp or Aesop shrimp!

To introduce yet another twist, in the US the word prawn is hardly ever used. Instead, the term shrimp covers a size ranging from what Brits would call shrimps up to 25 cm long whoppers from the Gulf of Mexico. This tendency is becoming more common in the UK where you will often see references to 'jumbo shrimp', mainly from the Far East.

With the increasing international nature of food supplies, and with supermarket fish counters offering fish and shellfish from all parts of the world, a much more important distinction is between the two major groups, not based on size, into which crustacean experts divide the shrimps and prawns. One of these, the *carideans,* consists of slow-growing, cold-water species, including the traditional British shrimps and prawns dealt with above. Even the biggest grow to no more than about 10 or 11 cm long and can take several years to do so. The second group, the *penaeids,* are mostly much faster growing warm water species that can reach monstrous lengths of almost 30 cm. Because of their fast growth, often reaching a marketable length in a matter of months, these are the ones that are naturally favoured by shrimp farmers and dominate the fish counters in most western supermarkets. They also seem to be popular with the consumers, presumably partly because of their size, but also because they tend to be less expensive.

Unfortunately, the fast growth rate goes along with a much inferior taste. Given the choice, I would go for carideans every time.

So apart from the price, how can you tell the difference between carideans and penaeids? If they are shelled it is quite difficult, but shell-on it's dead easy (see Fig 14). Like all ten-legged crustaceans, the tails of shrimps/prawns are made up of a series of (7) separate segments, each covered with thin shelly plates; these are the bits you peel off before you eat them. In carideans, the second tail segment has a big side plate that overlaps both the segment in front and the one behind. In contrast, the penaeids have all the side plates of the tail segments more or less the same size and all overlapping the one behind, a bit like the sections of a telescope. This may sound a touch technical, but it isn't, believe me. Try it, and then taste the difference!

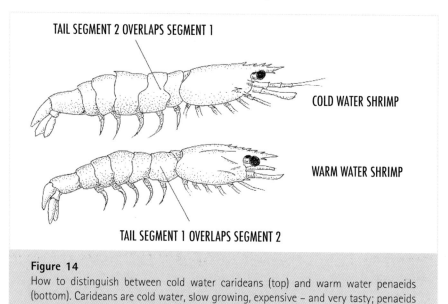

TAIL SEGMENT 2 OVERLAPS SEGMENT 1

COLD WATER SHRIMP

WARM WATER SHRIMP

TAIL SEGMENT 1 OVERLAPS SEGMENT 2

Figure 14
How to distinguish between cold water carideans (top) and warm water penaeids (bottom). Carideans are cold water, slow growing, expensive – and very tasty; penaeids are warm water, fast growing, cheap – and generally fairly tasteless!

47 What is or are krill?

Krill is a Norwegian word meaning young fish, more or less equivalent to the English word 'fry'. But it is also used internationally to refer to shrimp-like crustaceans that are crucial constituents of the diets of whalebone whales, manta rays, whale sharks, squid, some seals, and seabird species, including some of the penguins.

Krill belong to a group called the *Euphausiacea*, which contains about 90 species worldwide. They range in size as adults from a centimetre or so in total length up to about 15 cm. They occur in all oceans and at all depths, but most species live in the upper 500 metres or so and many live even closer to the surface where they feed by filtering out tiny phytoplankton cells. This is true of the most famous member of the group, the Antarctic krill, *Euphausia superba,* which grows to a length of 5-6 cm and occurs in huge quantities in the Southern Ocean where it is at the very centre of the Antarctic ecosystem.

E. superba congregate into vast swarms believed to contain up to 30,000 individuals per cubic metre, hence their attractiveness to whales, seals and penguins. Since krill feed directly on microscopic plants and are, in turn, eaten by the blue whale, the largest animal on earth, this is the most efficient food chain on the planet.

The total Antarctic biomass of krill at any one time is estimated at 500 million tonnes, roughly twice the weight of all the humans on earth. About half of this is eaten each year by whales, seals, birds, squid and fish, and is replaced by krill growth and reproduction, but krill are also exploited directly by man. They have been fished commercially since the early 1970s and the catches peaked at more than 400,000 tonnes in 1985-86. The annual catch has subsequently decreased to about 100,000 tonnes, with Japan and Russia being the main consuming nations.

Krill has been used for human consumption, but is has a strong and distinctive flavour and has proved not to be very popular. It also contains high levels of fluorine and is toxic unless the shells are completely removed. Finally, it has extremely powerful protein-digesting enzymes and becomes unfit for human consumption if not processed within three hours of being landed on deck. Consequently, it is now mainly used as a protein-rich additive to pig and cattle food.

Fish

48 How many different types of fish live in the oceans?

Roughly 12,000, that is, about half of the 25,000 or so fish species known from all the aquatic environments on earth[23]. Most marine fish species live over the

23   Freshwater fish faunas vary hugely in species richness. The mighty Amazon and its more than 1,000 tributaries provide homes for more than 1,700 described fish species (including about 20 piranhas), though there are probably still another 1,000 species so far undiscovered. The Mississippi system, with about one tenth of the water flow of the Amazon, has a correspondingly less rich fish fauna with only about 250 species. In contrast, the rivers and lakes of the British Isles contain a total of only 38 fish species!

relatively shallow continental shelves in depths less than about 200 metres. And about a quarter of all the species are associated with tropical coral reefs, ranging from the mighty groupers weighing in at up to almost half a tonne to tiny species no more than a centimetre long as adults. On the other hand, only 2,000 or so fish species have been described from the open oceans beyond the continental shelves. About 250 of these are from the near surface, 100 to 200 metres, and the remainder from deeper in the water column, with 350 or so different species living on or close to the sea floor. Although quite a few deep sea fish probably remain to be discovered, it is unlikely that these figures will change very substantially.

Considering the huge size of the oceans relative to the earth's freshwater lakes and rivers[24], and the fact that the oceans overlying the continental shelves contain only about 1/700th of the volume of the open oceans, the deep sea is clearly a very species-poor environment, at least as far as fish are concerned. Why should this be? Well, the principal reason seems to be that a vast amount of the deep ocean is fairly monotonous, with rather minor variations in such features as temperature and salinity. Furthermore, whereas the world's fresh-water habitats are broken up into innumerable streams, rivers and lakes, separated by inhospitable (for most aquatic animals) land areas, there are no such barriers in the open ocean, which might isolate different fish populations and stop them mixing and interbreeding. These two features, environmental variability and the possibility of at least some isolation between populations, seem to be key elements in producing lots of species, that is, high biodiversity. And coral reefs seem to provide both features in abundance, with lots of different foods and ways for fish to make a living, and lots of nooks and crannies for them to hide in. Different coral reefs, separated by inhospitable open water, can also act a bit like isolated islands, separating fish populations so that they can evolve in different directions without constantly interbreeding, and therefore blending, with each other. So that's why such a high proportion of marine fish are found on and around reefs.

But the same conditions favouring the development of lots of different species of fish affect other animal groups too, so coral reefs tend to be biodiversity 'hot spots' for everything from sponges and the corals themselves to worms, crustaceans and echinoderms.

In the same way, the absence of these biodiversity-friendly conditions in the open ocean have the opposite effect on all animal groups. So just as the deep water column has a relatively small number of fish species, most other

---

24   The oceans contain about 25 times as much water as all the rivers, lakes and ice caps put together.

animal groups are also represented by rather few species in the deep ocean – except, that is, for the sea floor. As Q31 explains, fifty or so years ago the deep seabed environment, like the overlying water column, was thought to be fairly monotonous and accordingly rather species poor. It still is, as far as large and quite mobile animals like fish and the larger crustaceans and echinoderms are concerned. After all, this environment represents well over half the surface of the earth but supports less than 2% of the planet's fish species. But at scales of a few cm to a metre or two, fairly irrelevant to a big mobile animal like a fish but representing its whole world to a tiny beast no more than a millimetre or two long, the sea floor seems to be extremely variable both in space and in time, just the features that encourage the evolution of lots of species. As a result, this amazing environment manages to be both a low biodiversity one (the deep water column) and, at the same time, possibly the most species-rich one on earth.

49 Are all fish species cold blooded?

No. Most are, but a few, notably tuna, can maintain their body temperature up to 10 or 20° warmer than the surrounding seawater. Generating heat for a fish is no problem because it is an inevitable by-product of muscular contraction. But for most fish this heat is rapidly lost through the gills where the blood in capillaries is exposed to the cold surrounding water. For many fish species this doesn't matter too much because their white muscles work acceptably well at low temperatures. But for tuna and some of their relatives the red muscle involved in fast swimming seems to work more efficiently at elevated temperatures. These fish therefore maintain the red muscle temperature in colder external conditions by using a counter-current heat exchange system. Warm blood flowing from the red muscle back to the gills to pick up more oxygen passes through a system of capillaries running very close to a similar system carrying oxygenated, but cool, blood from the gills to the muscle. Heat is transferred from the warm outflowing blood to the cold inflowing blood so that this arrives at the red muscle already somewhat warmed up. It picks up even more heat from the contracting muscle and, in its turn, passes some of this heat to inflowing blood as it flows back to the gills.

50 Which is the largest fish in the oceans?

The largest fish alive today is the whale shark, *Rhynchodon typus*. The largest specimen accurately recorded was 12.65 metres (41.5 ft) long and weighed more than 21 tonnes, but there are unsubstantiated claims of specimens more than 20 metres long.

The whale shark is an inhabitant of tropical and subtropical waters throughout the world oceans between about 30° north and south of the equator. It is a filter feeder, periodically taking a large volume of water into its huge mouth and expelling it through its gills, trapping any particles more than 2 or 3 mm in diameter on small outgrowths lining the pharynx and gill bars. The diet therefore consists of small pelagic animals including shrimps, squid and fish. This feeding technique is quite different from that of the basking shark, another filter feeder, which swims along with its mouth agape, relying on its forward momentum to force water through its gills.

Since the whale shark's food, like that of whales and dolphins, is most abundant in the upper layers of the oceans, the sharks generally stay close to the surface. However, specimens tagged with pressure sensors have been found to dive as deep as 1,000 metres.

Whale sharks are believed to reach sexual maturity at about 30 years old and females give birth to as many as 300 live 'pups' at a time, each about half a metre long. They are long-lived and possibly reach an age of 100.

They are, of course, totally harmless to man.

51 How do flying fish fly?

They don't actually fly, but simply glide using their fins like a glider's wings.

Like most other fish, flying fish have a series of unpaired fins, including the caudal or tail fin, and two sets of paired fins, the pectoral fins on the side of the body near the front end and the pelvic fins on the belly about halfway back. In a typical fish, like a herring for example, the upper and lower lobes of the tail fin are roughly the same size and the whole fin is used to drive the fish forwards. The pectoral and pelvic fins are also reasonably similar in size and are used mainly to steer the fish up, down and sideways. But in flying fish the pectoral fins are always hugely expanded to form 'wings', while the lower lobe of the tail fin is much longer than the upper lobe and is used as the 'engine' to launch the fish through the sea surface. In some flying fish the pelvic fins are also expanded so that in 'flight' they look a bit like a biplane with two pairs of wings.

But whether they have one pair of 'wings' or two, the same principles apply. While under water, the enlarged fins cannot be spread fully; if they were, they would impede conventional swimming very seriously. Instead, they are kept almost fully folded close against the side of the fish's body. But if the fish is scared by a predator – or a ship – it heads for the surface at full speed. Vibrating its caudal fin from side to side at several times a second, the fish accelerates to a speed of 5-10 metres a second, that is 10 to 20 knots, over a horizontal distance of no more than 2 or 3 metres, a phenomenal efficiency. As it breaks through the surface and leaves the water the 'wings' are spread and, as they move through the air, provide 'lift', just like a glider's wings, supporting the fish a few inches to a foot or two above the surface of the sea. If the fish has by chance headed into the wind, the air speed across the 'wings' may be considerably greater than the fish's speed across the water surface. Consequently, it may generate enough lift to carry the fish some metres upwards. This must be the reason they land on ships' decks far above the surface of the sea from time to time.

However, having left the liquid environment, the flying fish's engine, its tail fin, is now useless. Although it may still be vibrating from side to side, the air is far too 'thin' for it to provide any forward thrust. So sooner or later, and usually after only 3 or 4 seconds, the fish's weight will exceed the lift and it will fall back towards the surface. But if the sea conditions are suitable, particularly not too rough, the fish may have another bite of the cherry by vibrating its lower tail fin lobe as it dips into the sea and taking off for a second time. It may even do the same thing a third time, but this seems to be just about the limit. Each burst of tail-vibrating activity must use quite a lot of energy and after three successive bursts the fish's energy reserves seem to be pretty well spent. Nevertheless, a single phase flight may carry a flying fish 30 or 40 metres, and a three phase flight might take it well over 100 metres, a pretty safe distance from the original potential predator (or cruise ship) – as long as, by chance, it doesn't land just in front of the mouth of another hungry predator!

52 Why do flying fish fly?

Many fish leap out of the water from time to time, but the acknowledged champions are the true flying fish. There are about 60 species worldwide and all belong to the family *Exocoetidae*, a name derived from the Greek words *exo* and *koitos* meaning lying down or sleeping outside[25]. This name apparently derives from their unfortunate, and terminal, habit of landing from time to time on the deck of a ship as a result of their flying activities during the night, to be found in the early morning light – and usually promptly cooked and eaten! Adults range in total length from about 11 cm up to about 40 cm, though juveniles can, of course, be much smaller and sometimes look a bit like dragonflies. They occur mainly in tropical and subtropical seas, particularly between about 40°N and 40°S. They can often be seen from cruise ships sailing through the intense blue oceanic waters in these regions, usually flying away from the front of the ship, sometimes singly, but more often in groups of 10 to 20 or more.

Flying fish live constantly very close to the surface in clear, sunlit water containing very little food. They make a living by feeding on other small fish and planktonic (see Q26) animals, particularly when these are concentrated close to the surface during their daily migrations. But there are other beasts living in these surface waters, such as sharks, dolphins, tuna and so on, for which a flying fish would provide a tasty and much needed meal. In fact, predation pressure in these regions seems to be so high that it has resulted in the adoption (i.e. evolution) of an extreme escape technique, which is exactly what the flying fish flight is.

They fly in response to a perceived threat, usually from a potential predator, with the intention of getting as far away as possible in the shortest possible time. And as far as they are concerned, a ship is a huge and extremely noisy potential predator. After all, an observant cruise ship passenger may see hundreds of flying fish in the course of a few hours. But remember that it is quite possible, indeed very likely, that in the vast expanse of the oceans those flying fish had never before encountered a ship. No wonder they try to get away!

53 Do fish drink water?

It depends on the fish – and on the water! The cartilaginous fish, that is the sharks, skates and rays, are a bit special, so let's leave them to one side at the moment. We will deal first with the other major fish group, the bony fish, including all the 'ordinary' fish like cod, tuna, herring and plaice in the sea, and trout, pike, perch and so on in fresh water.

Like us, fish need water and a variety of mineral salts to keep them healthy.

---

25   Sadly, the same derivation was used for the French anti-ship missile, Exocet, that was used to such devastating effect during the Falklands war and the Iran-Iraq war.

Whether they live in fresh water or in the sea, fish are surrounded by plenty of water, but the salt situation couldn't be more different. In fresh water, salts are in very short supply, while marine fish are confronted with an embarrassment of salty riches, so both face problems with osmosis. Osmosis is that strange phenomenon that most of us were introduced to at school, whereby if two solutions of different strengths are separated by a 'semi-permeable membrane', water moves across the membrane from the weaker to the stronger solution until they are both of the same strength[26]. There are lots of semi-permeable membranes in nature including, for example, our gut lining and fish gills.

The internal fluids of fish, particularly the blood, contains much more salt than fresh water and much less than seawater (about 0.4% salt for freshwater fish compared with virtually zero for freshwater, and 1.4% salt for the blood of marine fish compared with about 3.5% for seawater). Consequently, water moves consistently across the gills *into* fish in freshwater, but brings virtually nothing by way of mineral salts with it. Drinking would simply make matters worse, so freshwater fish generally do not drink, though they can't help swallowing at least some water as they feed. Their kidneys therefore have to filter the blood, retaining the salts, and getting rid of the water. Freshwater fish therefore produce large quantities of very dilute urine, so that I like to think of rivers and lakes as being basically dilute fish pee!

---

26   If you find this concept a bit difficult to grasp, think of the semi-permeable membrane as a sheet of chicken wire stretched as a barrier across a container filled with lots of balls, some bigger than the holes in the chicken wire, like tennis balls, and some much smaller than the holes, say like ping-pong balls. These balls represent the molecules making up the solutions on either side of the semi-permeable membrane, the small balls representing the relatively small water molecules (the solvent) and the big ones representing the various chemicals dissolved in the water (known as the solutes). Now imagine that although there is the same total number of balls on either side of the barrier, on one side there are lots of small balls, but also lots of big balls, representing a strong solution, while on the other side, representing the weak solution, most of the balls are small (i.e. 'water') with very few of the bigger balls. Finally, imagine huge fans blowing the balls in all directions a bit like lottery balls immediately before the draw.

Clearly, when the balls are blown against the chicken wire many of the small ones will go straight through because they are smaller than the holes, whereas the big balls will be stopped. Note that the small (water) balls coming up against the barrier will pass through it in either direction equally easily. However, at the beginning of our little experiment the 'weak solution' side of the barrier had many more small balls than the 'strong' side. So at first, most of the balls passing through the wire mesh would be from the 'weak' side to the 'strong' side, with rather few coming in the opposite direction. But this difference would get smaller and smaller as more and more 'water' balls move into the 'strong solution' side, effectively making it weaker by increasing the number of solvent molecules relative to the solute molecules. Theoretically, the difference should disappear completely (so that 'water' balls cross the barrier in both directions at the same rate) when the ratio of small balls to big balls is the same on both sides. This is exactly what happens in nature. When the fluids on either side of a semi-permeable membrane reach the same strength (called the osmotic strength), the flow of solvent molecules is the same in both directions.

The reverse problem is faced by fish in the sea. Here, water tends to leave the fish via its gills and its gut lining and this loss has to be made up if the fish is to avoid dehydration. So marine fish do drink, sometimes as much as a tenth of their body weight each day. But this poses another problem. Their kidneys, like ours, have to get rid of toxic chemicals that are inevitably produced during normal metabolism, but since they can't afford to lose water they produce small amounts of very strong urine including most of the toxins. However, this can't eliminate all the salts that came in with the water. Instead, the salts are pumped out, against the concentration gradient, across the gills, a process that uses quite a lot of energy.

Some fish, of course, occur in both environments, either switching backwards and forwards throughout their lives, or spending part of their life history in one environment and the rest in the other. Fish like sticklebacks and flounders, frequently found in estuarine environments, belong to the first group and seem to be able to switch from the fresh water to the marine fish role easily and rapidly. Salmon and eels, on the other hand, tend to make the switch only once and seem to find changing back difficult, if not impossible. Fish like eels, which spend most of their adult life in fresh water and migrate down to the sea only towards the end of their lives in order to breed, are known as *catadromous*, from two Greek words that simply mean 'to run down' (i.e. down river). Salmon, on the other hand, which spend most of their adult life in the sea and only return to rivers in order to breed, after which most quickly die, are called *anadromous,* meaning, would you believe, 'to run up'! Both groups have to cope with a major change in the saltiness of their surroundings and seem to do so by switching part of their kidney function on or off, as appropriate.

The bit involved is the glomerulus, a ball of tiny blood vessels inside an egg-cup shaped structure called Bowman's capsule, the two parts together making up a renal capsule. Most vertebrate kidneys have lots of renal capsules which allow water and unwanted chemicals to leave the bloodstream and be passed into the urine. Clearly, the renal capsules of freshwater fish need to work overtime, while those of marine fish have a relatively easy job. The trick that eels and salmon have to achieve is to switch from one to the other. This is possibly why salmon gather for some time near the mouth of their chosen river before swimming into it, perhaps to give their kidneys time to adjust.

Finally, let's return to the cartilaginous fish, that is the sharks, skates and rays and their relatives. They have been around for a very long time, probably a good deal longer than their bony fish relatives. And over the course of evolution they have come up with a novel answer to the water problem. Most cartilaginous fish are, and always have been, marine, though a few wander into fresh water and an even smaller number live throughout their lives in fresh water. All of them have blood with a salt content very similar to that of marine bony fish, that is about half as salty as seawater. However, instead of losing water by osmosis as the bony fish do, the marine cartilaginous fish actually gain a little bit of water which they can get rid of fairly easily via the kidneys. To do this they have a clever trick up their sleeves, or rather in their blood vessels.

In addition to the normal blood salts, cartilaginous fish blood contains high levels of nitrogen-containing substances, the main one being urea. Most vertebrates make urea as a waste product and a way to get rid of toxic nitrogen containing materials, resulting mainly from the breakdown of proteins. Not surprisingly, for most vertebrates, including bony fish, high concentrations of urea and similar substances would be very toxic – but not for the cartilaginous fish. They are very tolerant of these nitrogenous materials – and understandably so, because, along with the normal salts, they bump up the osmolarity of their blood to a little higher than that of seawater, thus reducing the salt and water balancing problem to very manageable proportions. When they die, of course, the urea is still in the blood and tissues and would make them very unpalatable to humans. That's why large skate were traditionally towed by fishermen behind the boat as they came back to port to allow the breakdown products of urea, including ammonia, to dissipate before they came to market.

Finally, what about those few cartilaginous fish that live in fresh water? Well, it seems that they produce and retain much less urea than their marine relatives and therefore have blood with a much lower osmolarity. However, this means that they have much the same problem with water and salt as freshwater bony fish so, like them, freshwater sharks have to drink and pee a lot!

Birds

54 Why do seabirds follow ships?

Probably for at least three different reasons: as a potential source of food; to hitch a lift; or simply out of what we would call curiosity.

The food source explanation probably has two origins. First, many inshore and shallow feeding birds, including gulls, terns, gannets and boobies, probably learn at an early age that fishing vessels are good sources of food as fish escape from the nets during hauling and the fishermen subsequently gut the fish and throw the trash overboard. In the birds' minds any ship therefore probably represents a possible food bonanza. But all ships, and particularly big ones, also provide potential food as small crustaceans and fish lurking a few metres down are brought to the surface in the turbulent waters of the ship's wake. Terns are often seen taking advantage of this phenomenon, fluttering over the wake just above the surface and apparently paddling along on their feet and occasionally dipping their bills into the water.

For many long distance gliders, such as boobies, gannets, frigate birds and, particularly, albatrosses and their tube-nose relatives (see Q55), ships in the middle of the ocean are welcome sources of heat and, as a result, rising air thermals that the gliding birds can exploit just as a human glider pilot does. Ocean going ships, including cruise ships, are from time to time surrounded by such birds, sometimes dozens of them, wheeling all around the vessel for hours or even days on end. And they can disappear just as suddenly as they arrived. Who can say whether their presence is because the ship happens to be going in roughly the same direction as the birds at the time, or is simply an expression of their curiosity and sense of fun?

But hitching a lift can mean exactly that, when distressed birds land on ships in a desperate last ditch effort to survive, usually during the day. These boarders are often not seabirds, but land birds exhausted by bad weather on their normal migration route or, more often, blown off course and out to sea. Probably the best thing you can do for these reluctant hitchhikers is to offer them some fresh water, something they have been unable to get while they have been over the sea. Sadly, however, they rarely survive the trauma they have experienced.

In contrast, night-time avian visitors often have a much better chance of survival. Many of them are true seabirds, like petrels and shearwaters. These are maritime professionals, quite capable of surviving at sea, but have landed on the ship because they have been disorientated in the darkness by the ships' lights. Assuming they haven't damaged themselves by crashing into a lamp or part of the ship's superstructure, these birds have a good chance of survival and, after a bit of a rest, they often leave the ship and carry on with their lives.

55 How many different seabirds are there?

It depends a bit on what you consider to be seabirds, but one of the standard works for maritime birdwatchers[27] lists some 320 different species and forms, of which about 45 are regularly seen in north Atlantic/British waters.

This book does not pretend to be a guide for birdwatchers at sea; it simply

---

27 Peter Harrison's *Seabirds of the World; a photographic guide*, originally published in 1987 and reprinted several times. My copy is a paperback published by Princeton University Press in 2003. It is a very accessible handbook based on the much more definitive, and 220,000 word long, *Seabirds: An Identification Guide*, also by Peter Harrison and published by Beckenham in 1983.

tries to answer some of the questions that non birdwatchers often ask. However, it will hopefully encourage a few of the latter to take up birdwatching on cruises, even if not too seriously, because it can provide a huge amount of pleasure. Not only are seabirds fascinating in their own right, but they can also tell you a lot about what is going on beneath the ocean surface. In general, no birds about means no food, so not much chance of seeing anything else like whales, dolphins or seals. On the other hand, lots of birds feeding in the same patch of water means that other beasts, like whales and dolphins, are likely to be attracted to the same spot – so keep an eye open for them!

The classification of seabirds is a matter of some controversy among ornithologists, but most of them are put into one or other of four main groups called Orders, each with a rather strange scientific name ending in ... *formes*.

The *Sphenisciformes* (from the Greek word *sphen* meaning wedge) contains only the rather wedge-shaped penguins.

The *Procellariiformes* contains all the really oceanic birds like the albatrosses and the fulmars, petrels and shearwaters, about 100 species in all. The name is derived from the Latin word *procella,* meaning a violent wind or storm, referring to the apparent ease, and even pleasure, with which these ocean wanderers seem to deal with bad weather. They are also often collectively referred to as tube noses because, while the albatrosses have single nostrils on either side of the hooked bill, the others all have the nostrils united into a single prominent tube on the top of the bill. The nostrils in all the *procellariiforms* act as conduits for a very salty discharge which is the mechanism by which they get rid of the excess salt that they inevitably take in as they feed. So if they come close enough, you can often see a salty dewdrop on the end of the beak.

The *Pelecaniformes*, with about 40 species, contains (you guessed it) the pelicans, but it also includes the cormorants and shags, the frigate birds and tropic birds, as well as the gannets and their tropical relatives, the boobies. Most *pelecaniforms* are spectacular divers, and the splashes of a group of gannets or boobies entering the water can be seen from a long distance away and can be a good indication of the presence of other hunters like whales and dolphins.

Finally, the *Charadriiformes* contains the rest, well over 100 species dominated by the so-called seagulls (see Q56), terns and noddies. But the *charadriiforms* also includes the skuas, the aggressive bruisers of the seabird world, and the guillemots, razorbills and puffins, grouped together as the auks, which are superb underwater swimmers. The auks are confined to the northern hemisphere, where they occupy more or less the same ecological role as the penguins in the southern hemisphere. Unlike the penguins, however, the auks can fly, albeit not nearly as efficiently as many other birds.

56 What are seagulls?

Seagulls, more correctly simply gulls, are members of the bird family *Laridae* which also contains the terns and noddies. The four noddy species are all tropical or subtropical, but the 37 tern species and the 47 gulls are distributed all over the planet from the Arctic to the Antarctic. In fact, the Arctic Tern breeds during the northern summer in the Arctic and subarctic regions of Europe, Asia and North America, and overwinters in the southern hemisphere, some individuals even reaching the islands around Antarctica. This represents an annual migration of well over 20,000 miles (32,000 km), the longest regular migration of any known animal.

The true gulls are generally much less migratory and some do not migrate at all. But many northern hemisphere gulls, including north-western Europe's most familiar large gull (the Herring gull) and small gull (the Black-headed gull) spend the winter around the Mediterranean and North Africa and India.

Although gulls are associated in our psyche with the seaside, they are not true seabirds in the sense that they are not confined to a maritime lifestyle. In fact, the largest UK population of herring gulls in 2006 was said to be on the municipal tip at Gloucester, hardly a typical seaside locality. Similarly, as I write this piece on the first floor of my house in Alton in Hampshire I am looking out over the local football pitch where the most abundant birds are black-headed gulls, despite the fact that we are some thirty miles from the sea. Such a lifestyle is to be expected from a gull, because these birds are extremely efficient generalist and opportunist feeders, able to exploit virtually anything remotely edible.

57 How many penguin species are there?

Probably 18, though one of these, the White-flippered Penguin found only near Canterbury, New Zealand, is usually considered to be a subspecies of the much more widespread Little Penguin found throughout New Zealand and southern Australia.

The White-flippered and Little Penguin (also known as the Blue Penguin) are the smallest species, standing only about 40 cm or 16 inches high. At the other extreme, the largest penguin species is the Emperor, standing 112 cm or 44 inches high!

58 How deep do penguins dive?

Like whales, dolphins and seals, penguins dive mostly in search of food. Since their food, consisting mainly of krill, fish and squid, is generally concentrated fairly close to the surface, they have no need to dive very deep, nor for very long. Consequently, most of the smaller species dive for periods of only two or three minutes at a time and to depths of no more than 10 or 20 metres. However, they can probably dive considerably deeper if necessary, for example to escape predators like the killer whale or leopard seal. Moreover, some of the larger species probably dive routinely to greater depths and the Emperor Penguin is recorded as reaching depths of more than 500 metres and staying submerged for more than 20 minutes.

59 Do all penguins live in the Antarctic?

No! In fact, only five of the 18 or so recognised species occur on the Antarctic continent. Four of these, the Adélie, Macaroni, Gentoo and Chinstrap, breed during the southern summer, either on the coasts of Antarctica (the Adélie), or on the Antarctic Peninsula or the sub-Antarctic islands. The fifth Antarctic species, the Emperor Penguin, is the only one that breeds on the Antarctic continent during the southern winter. The males huddle together in hundreds or thousands for mutual warmth against temperatures as low as -30°C while incubating a single egg held above the ice on the penguin's feet and covered by a flap of skin[28].

The non-Antarctic penguins live around New Zealand and Australia, southern Africa and South America, where one species, the Humboldt Penguin extends as far north as Peru, only 5°S of the Equator.

Finally, the most northerly penguin species is the Galapagos Penguin which, as its name suggests, is restricted to the Galapagos Islands straddling the Equator some 500 nautical miles off the coast of Ecuador.

No penguins live in the northern hemisphere.

---

28 This amazing phenomenon was the subject of the critically acclaimed Warner Independent nature documentary film, *March of the Penguins*, originally written and directed by Luc Jacquet and Michel Fessler and released in French in 2005 as *La Marche de l'empereur*.

60 Where does the name penguin come from?

As you might guess, there are a number of theories.

My favourite is that it was derived from two Welsh (or possibly Breton) words, *pen* meaning head, and *gwyn* meaning white, the name *pengwyn* having been applied to the Great Auk, a flightless relative of modern day razorbills and puffins, which became extinct in the mid nineteenth century. The great auk was a large black and white bird, remarkably similar in general appearance to penguins, though not related to them, and was widespread in the North Atlantic from Canada in the west to the British Isles and Norway in the east. It was extensively exploited as food and for its down for hundreds of years. It would have been entirely reasonable for European mariners to apply the same name to penguins when they encountered them in the southern hemisphere.

A less attractive (at least to me) theory is that the name comes from the Latin word *pinguis* which means fat, plump or fertile, words that are certainly conjured up by the somewhat comely appearance of penguins.

Finally, an even less satisfactory theory is that it is a corruption of the term 'pen-wing' referring to the rudimentary limbs of both the Great Auk and penguins.

Take your pick!

61 Do penguins have feathers or fur?

Like all birds, penguins have feathers. But because they are not used for flight, penguin feathers are highly modified and superficially look a bit like fur. They are short, particularly on the wings, and are more closely and evenly spaced than in most other birds. The outer part of each feather is waterproof, while an inner, downy part traps a layer of air close to the penguin's skin as insulation against the cold.

This insulation is so efficient that most penguin species have more trouble keeping cool in warm air than keeping warm in cold water! Several of them nest in rookeries in South America, southern Africa and Australasia where air temperatures frequently reach the upper 20s centigrade during the breeding season. Under these conditions penguins often cool down by standing with their wings outstretched to radiate heat and fluff up their feathers to flush out some of the air trapped between them.

On the other hand, most penguin species spend at least some time in very cold waters or standing on snow and ice. The extreme example is the Emperor penguin which breeds in the Antarctic during the southern winter. Clearly, under these conditions good insulation from feathers and fat is essential. Apart from the beak and the eyes, the only parts of a penguin's body that are not covered in insulating feathers are the feet. So why don't they freeze solid in contact with the ice? In fact, they almost do, but not quite.

Penguins maintain their internal body temperature (and their blood) at about 40°C, not very different from ours. If they were to keep their feet at this temperature by pumping lots of warm blood into them, the resulting huge loss of heat would be difficult, if not impossible, to make up. Instead, in winter the feet are kept only a degree or so above freezing and much colder than the rest of the body. They do this partly by restricting how much blood goes into the feet by constricting the arterial vessels supplying them, but also by having a heat exchange system quite similar to that used by tuna to keep their bodies warmer than the surrounding water (see Q49). In penguins, the arteries supplying the feet are subdivided into finer and finer vessels which run alongside similarly fine venous vessels carrying blood back from the feet to the main body. Heat from the arterial blood is transferred to the neighbouring returning venous blood, so that the blood eventually reaching the feet is only 'lukewarm' – just enough to stop them freezing, but not enough to be a serious drain on the bird's heat reserves.

Whales and Dolphins

62 How many whale and dolphin species are there in the oceans?

There are 80 species of whales and dolphins recognised worldwide at the moment. They all belong to the mammalian order *Cetacea* (simply meaning large sea animal) and range in size from small dolphin species just over a metre (39 inches) long to the blue whale, which can reach a length of 33 metres (110 feet) long and weigh as much as 190 tonnes, the largest creature that has ever existed on earth. Since they are mammals, as we are, cetaceans have warm blood and breathe air, just as we do. So they have to come to the surface periodically to breathe, which is when we are sometimes lucky enough to see them, but also when some of us are able to hunt them.

Although it is possible that additional species will be discovered or recognised in the future, some species, such as the Yangtze river dolphin or baiji, are under threat or possibly already extinct, so that the total number of whale/dolphin species is unlikely to change significantly.

63 How deep do whales dive?

In general, the bigger the whale, the deeper the dive. However, the deepest diver of all is the Sperm whale which, at a maximum length of 'only' about 16 metres or 60 feet, is by no means the biggest. Sperm whales can stay submerged for as much as two hours and seem regularly to reach depths of 1,000 metres or more. In fact there is some evidence that they may dive as deep as 3,000 metres or 10,000 feet! As for all whales, the reason Sperm whales dive is to find food, in their case particularly large squid. And since these squid live deep in the ocean, the whales have to dive deep too.

But the food of the other whales and dolphins lives generally much shallower. The whalebone whales, including the very biggest rorqual species like the fin whale and the blue whale, feed by sieving vast volumes of water through their baleen, or whalebone, plates (see Q75). They feed on huge shoals of small fish and crustaceans, particularly the shrimp-like krill (see Q47). Since these, in turn, are dependent upon the plant growth in the near surface waters they are mainly confined to the upper 250 metres or so, and this is about as deep as the big whales dive. Dives usually last for no more than 10 or 20 minutes, although several species can stay down longer than this and humpbacks sometimes remain submerged for 45 minutes or more.

The fast swimming and predatory toothed whales and dolphins feed on larger individual fish, crustaceans and squid and stay even closer to the surface, generally diving no deeper than a few tens of metres and staying submerged for only five to ten minutes. Even the most voracious and largest of the dolphins, the killer whale, feeding on almost anything from squid, fish and birds to seals, turtles, dolphins and even young rorquals, probably dives no deeper than about 100 metres, though they can stay submerged for 15 minutes or more.

64 Do all whales and dolphins come to the surface?

Yes. As Q62 explained, the group *Cetacea*, which contains all the whales, dolphins and porpoises, are mammals, just like us. And this means that, like us, they are warm blooded, they suckle their young on mother's milk, and they have lungs and breathe air. So, unlike fish, they cannot obtain their oxygen from seawater flowing over their gills. Instead, they must come up to the surface periodically to replenish their air supply. Although some of the bigger species, particularly the deep diving sperm whale (see Q69), can stay beneath the surface for quite long periods, possibly occasionally for 2 hours or more, even they have to return to the surface at least 10 or 20 times a day. Many species, however, dive for much shorter periods, usually no more than 10 or 20 minutes. Clearly, these species surface 100 times or more each day.

And this answers another frequently asked question; do whales and dolphins prefer or avoid any particular time of day or weather conditions to come to the surface? No, they have no choice in the matter. Since they have to come to the surface to survive, they have to do so at night and during the day, and no matter how rough the sea is.

65 Do whales have to breathe out like scuba divers when they surface?

No, because a whale's lungs (or those of dolphins and seals) can never become 'over-filled' like those of a scuba diver. As explained in Q116, a scuba diver breathes air from his air tanks at the same pressure as the surrounding water. In contrast, a whale fills its lungs with air at the surface, just like a snorkel diver. For both of them, their lungs become squeezed by the water pressure as they dive deeper and deeper in the sea. So at a depth of about 10 metres, where the pressure is twice that at the surface (see Q7), the lungs of the whale and snorkeler will be only half their volume at the surface. Another 10 metres down, where the pressure is three times atmospheric, the lungs will be only one third the surface volume – and so on. When they come back to the surface, assuming they haven't released any air in the meantime, both the whale's lungs and the snorkeler's lungs will expand to exactly the same volume they had when they started the dive except for the amount of air, tiny in the case of a snorkel diver, absorbed into their bloodstreams while they were beneath the surface.

66 Do whales get the bends?

The bends, or decompression sickness, is a condition suffered by scuba divers if they surface too quickly. It occurs when nitrogen dissolved in the blood and body tissues comes out of solution as the pressure decreases. The process is similar to what happens when you open the cap of a bottle of sparkling water and bubbles of gas, in this case carbon dioxide, fizz out of the water. In the unfortunate scuba diver, the nitrogen bubbles become trapped in the circulatory system or in the muscles, often in the joints, causing serious pain and cramps (hence the name 'bends'). The condition is always painful, but can be very serious, or even fatal if the bubbles become lodged in the brain. Divers can reduce the risk of bends by either restricting the length of their deep dives or by making a series of stops during their ascent[29] to allow the nitrogen to leave their bloodstream slowly and be breathed out via the lungs[30].

The problem is peculiar to scuba divers and is unknown in snorkel divers. This is because the scuba diver's Aqualung supplies him (or her) with air at the pressure of the surrounding water, no matter what the depth (see Q116). Consequently, each time he breathes in, his lungs fill with air, almost 80% of which is nitrogen, thus providing plenty to dissolve in the bloodstream. In contrast, the snorkel diver fills his lungs at the surface with air at a pressure of one atmosphere or about 15 pounds per square inch. Having left the surface the snorkel diver has no further access to air, so his lungs are squeezed by the increasing pressure as he descends (see Q7) to one half the original volume at 10 metres depth, one third the volume at 20 metres and one quarter at 30 metres[31]. One effect is that the area for the absorption of nitrogen (and for that matter oxygen) into the bloodstream is greatly reduced. Coupled with the relatively short duration of snorkel dives, no more than a minute or so for most of us and five or six minutes for the extreme sport enthusiasts, the result is that insufficient nitrogen is dissolved into a snorkel diver's blood to cause any problem during ascent.

Whales and dolphins fill their lungs at the surface, just like snorkel divers, so they also start off with a limited amount of bend-inducing nitrogen. Furthermore, as whales dive deeper and their lungs are compressed, much of the contained air is forced into the windpipe and the extensive nasal passages, which have rather thick linings that decrease the passage of gases, including nitrogen, into the tissues.

But dolphins and whales dive for much longer periods than snorkel divers, up to 2 hours or more in the case of sperm whales, and such long dives would probably allow significant amounts of nitrogen to dissolve in the blood and

---

29 For example, when scuba instructor Mark Ellyatt broke the world scuba depth record in 2003 and dived to 313 metres (1,027 feet) off Phuket in Thailand, his descent took only 12 minutes, but his return to the surface took almost 7 hours.

30 Or by using a gas mixture made up of oxygen and a replacement for nitrogen, usually helium, which is less soluble in blood and therefore avoids the problem.

31 The world record for a dive based on one lungful of air is an almost unbelievable 214 metres made by Herbert Nichst in 1997. At this depth Mr Nichst's lungs would have occupied rather less than one twentieth of their volume at the surface! The space would have been occupied by his abdominal organs pushing upwards against his diaphragm. At the time, his tummy would have looked extremely concave and rather uncomfortable!

tissues despite the limited source. Until recently, whale biologists believed that a combination of the anatomical adaptations and behaviour, in which deep diving whales return to the surface slowly enough to allow the dissolved nitrogen to be dissipated into the lungs just like a slowly ascending scuba diver, meant that they did not suffer from the bends. This seemed reasonable since, after all, cetaceans have been about for some 50 million years so that they should have had plenty of time to evolve immunity from this potential problem of living in their chosen environment.

However, in recent years some indirect evidence has come to light that suggests that this may not be strictly true. It comes from the examination of tissues of stranded whales and the bones of long dead whales held in museum collections. In both sets of samples, the pathology suggests that the whales had actually suffered from bends, causing local changes in the bone structure. In the case of the recent specimens it seems that the animals might have been frightened into surfacing far too rapidly, possibly scared by human activities. But in the case of the 'old' bones it seems that their original owners were also quite old and that their bends symptoms were chronic rather than acute and had resulted from the accumulation of small effects over many years.

So the answer to the original question 'Do whales get the bends?' seems to be 'Not sufficiently to affect them in the short term, but cumulative small effects may make old whales a touch stiff'. Sounds a bit like humans and arthritis, doesn't it?

67 What's in a whale's blow?

The cloud of 'steam' issuing through the sea surface when a big whale comes up to breathe is about all we usually see before the huge beast disappears once more beneath the waves. It is clearly not really steam, but what exactly is it that makes it so obvious?

It's a bit of a mystery, but it seems to consist mainly of tiny water droplets which look white, just like steam does. Some of these water particles have condensed out of the whale's warm breath when it comes into contact with colder air, just as our own breath condenses on a cold winter morning. But some of them come from the small amount of seawater accumulated in the blowhole or holes on the top of the whale's head and which is violently expelled in the huge 'cough' as it empties its lungs of the old, stale air and immediately refills them with fresh air.

Finally, it seems that the explosive exhalation also ejects mucus from inside the whale's nasal passages and lungs, a bit like a gigantic sneeze!

68 Why do the blows of different whales look different?

The whale's blow is produced as it explosively empties its lungs when it comes to the surface. The blow mainly consists of a cloud of tiny water particles (see Q67), but its shape and size depends on the form of the whale's nasal passages and the blowhole (its nostril). And if it is very windy, of course, any distinctive features will be rapidly obliterated.

All whales have two nasal passages, both opening separately in all the whalebone whales (see Q75). Consequently, in some species, notably the right whales (see Q73), the blow is distinctly V-shaped, often with one limb clearly bigger than the other. In all the rorquals (see Q74) the blows from the two nostrils combine into a single jet, though it may look a little asymmetrical. In the Humpback the blow is very distinctive and bushy, being as wide as or wider than its height. In contrast, the blows of Fin and Blue whales are much narrower relative to their height and therefore tend to look taller.

In all the toothed whales, that is all the dolphins, including the Killer whale, all the Pilot whales and the Sperm whales, there is only a single external opening so that the resulting blow is never divided. In fact, the blow of the smaller species is small and indistinct and rarely visible except in very cold air. But the blow of the Sperm whale is probably the most distinct of all because its single nostril is slit-like and placed slightly to the left-handside of the front of its head. Consequently, the blow is directed at an angle to the front and to the left of the whale. Of course, from some viewpoints this blow will look vertical, just like those of other whales. But if you see a blow that leans over to one side in calm weather you can be fairly confident that you are in the company of a Moby Dick!

69 Why is the Sperm whale called a Sperm whale?

Apparently because the very fine whitish waxy material filling the Sperm whale's melon, that is the big space in the animal's forehead, was originally mistaken for sperm and named *spermaceti*[32] accordingly. I can do no better than give the following quotation from Herman Melville's *Moby Dick*, the most famous Sperm whale of them all:

---

32 The word *spermaceti* simply means whale sperm.

'This whale, among the English of old vaguely known as Trumpa whale, and the Physeter whale, and the Anvil Headed whale, is the present Cachalot of the French, the Pottfisch of the Germans, and the Macrocephalus of the Long Words . . . It is chiefly with his name that I now have to do. Philologically considered, it is absurd. Some centuries ago, when the sperm whale was almost wholly unknown in his proper individuality, and when his oil was only accidentally obtained from the stranded fish; in those days spermaceti, it would seem, was popularly supposed to be derived from a creature identical with the one then known in England as the Greenland or Right whale. It was the idea also, that this same spermaceti was that quickening humor of the Greenland Whale which the first syllable of the word literally expresses. In those times, also, spermaceti was exceedingly scarce, not being used for light, but only as an ointment and medicament. It was only to be had from the druggists as you nowadays buy an ounce of rhubarb. When, as I opine, in the course of time, the true nature of spermaceti became known, its original name was still retained by the dealers; no doubt to enhance its value by a notion so strangely significant of its scarcity. And so the appellation must at last have come to be bestowed upon the whale from which this spermaceti was really derived.'

Melville's reference to 'the Long Words' is to the creature's scientific name, at the time he was writing (1851), *Physeter macrocephalus* (the large-headed physeter), given to it by the father of modern taxonomy, Carl Linnaeus, in his compendium of all animal species known to him, *Systema Naturae*, published in 1758. Unfortunately, Linnaeus' knowledge of Sperm whales was second-hand and largely based on hearsay. On the basis of the quite different accounts he heard or read, he decided (wrongly, as it turned out) that there were four different species, to each of which he gave a separate name.

The first specific name on the list was *catodon*, but for some reason Linnaeus' second name, *macrocephalus*, became more or less universally adopted by scientists for many years. But there are strict and internationally agreed rules about the scientific naming of animal (and plant) species. Where, as in this case, several different names are given to a single species the rules are rather clear; in general the chronologically first name should have priority. And where different names are given in the same publication (as in *Systema Naturae*), and therefore appeared on the same date, the first one to appear in the publication (that is, nearest the front!) should win. In this case, the name *catodon* was unquestionably ahead of *macrocephalus*, and should therefore take priority. But taxonomists are generally a fairly reasonable lot and most of them think the rules are for guidance and not for blind adherence. So since the name *macrocephalus* had been more widely used than *catodon* for well over a century, the case has been made for retaining it despite its subservient position if the rules are applied strictly. Nevertheless, the jury is still out and you may well come across both names in the literature. But be aware that, just as there was only one Moby Dick, there is only one proper Sperm whale, though it has a couple of dwarf relatives.

70 What is the Sperm whale's spermaceti for?

Q69 described how the Sperm whale got this name because early whalers thought, rather amazingly, that the animal's bulbous forehead, or melon, contained sperm (despite the fact that female Sperm whales have a similar organ to the males, albeit a bit smaller). Although this idea was clearly nonsense, the true function of the melon and its contents, both unique to Sperm whales, is still the subject of some discussion among whale biologists.

One rather far-fetched suggestion is that it is used as a weapon by males in their competitive fights for the rights to females, presumably banging their heads together much as fighting deer stags bash their antlers together. A more plausible idea is that the wax or oil-filled melon acts as a sort of inbuilt loud haler system, concentrating and focusing the sounds produced by the whale to echo-locate its squid prey in the inky blackness of the deep sea. But an even more ingenious idea was put forward many years ago by a colleague of mine, Dr Malcolm Clarke, an expert in the interactions between squid and Sperm whales. Malcolm suggested that the Sperm whale uses its melon as a buoyancy aid during its long and deep dives in search of food. The spermaceti, or waxy material filling the melon, has a melting point at about 29°C, so at the whale's normal body temperature of about 33.5°C it is usually in its liquid state. However, the melon is not just a vast space filled with spermaceti, but is criss-crossed by a network of sinuses and nasal passages and is surrounded by blood vessels. So Malcolm Clarke suggested that, as the whale dives into colder and deeper water, the animal increases the flow of cold water through the melon causing the wax to solidify and shrink, thus increasing the density of the whale's head and helping it to sink.

Conversely, when the whale starts its ascent it increases the flow of warm blood through the head causing the wax to melt. In doing so, it increases its volume and becomes more buoyant, helping the whale to rise. Nice idea!

71 Are dolphins whales?

Yes, sort of.

Zoologists divide the *Cetacea*, the group to which the whales belong (see Q62), into a number of different sub-groups. The first great division is into the whalebone whales (or *Mysticetes*), the ones that feed on small animals they sieve out of the water through their whalebone or baleen plates (see Q75), and the toothed whales (the *Odontocetes*). The toothed whales, as the name suggests, have teeth instead of whalebone and feed on larger individual prey animals, particularly fish and squid. They range in size from the huge Sperm whale, at about 18 metres or 60 feet long and weighing in at up to 50 tonnes, to tiny dolphins no more than one metre long. They are divided into several families, the largest being the *Delphinidae* or true dolphins, containing 32 species. Most of them are no more than 2-3 metres long, but the family includes the awesome Killer whale or Orca, up to almost 10 metres (30 feet) long, and found throughout the world's oceans. Killer whales are voracious hunters, feeding on fish, squid, seals and even other whales. Like other dolphins, they are very intelligent, a feature that unfortunately makes them attractive to seaquarium operators who teach captive animals to perform tricks to order, surely a sad role for a magnificent animal.

72 What is the difference between a dolphin and a porpoise?

In some parts of the world, and particularly in North America, the name porpoise is applied to any small member of the *Cetacea*, the group to which whales and dolphins belong. But to zoologists there are just six species of true porpoises, all included in the family *Phocoenidae*, separating them from the dolphin family *Delphinidae* (see Q71). The two families differ in a number of anatomical details, including the fact that dolphin teeth are sharp and conical in shape while porpoise teeth are spade-shaped.

But the two families also differ in behaviour. Dolphins are usually very active, often leaping high out of the water and frequently approaching boats and ships, and they tend to live in groups of tens or even hundreds of animals. In contrast, porpoises tend to be very shy and slow moving, they live alone or in small groups, and they rarely show more of themselves than a brief glimpse of their backs as they surface to breathe.

There is just one porpoise species in British waters, the Harbour Porpoise which, as the name suggests, tends to be restricted to relatively shallow inshore waters, but is found throughout the north Atlantic and north Pacific. It seems that the Harbour Porpoise reminds some people of a pig, and an alternative name for it is the Puffing Pig. In fact, the name porpoise itself is apparently derived from the French words *porc poisson* meaning 'pig-fish'.

73 Why are Right whales called Right whales?

Sadly, it is basically because during the 17th, 18th and early 19th centuries they were the *right* whales to hunt.

Before the widespread introduction of steam engines into ships, essentially from the middle of the 1800s, the only way to get close enough to a whale to harpoon it was in a rowing boat launched from a sailing mother ship. Once killed, the whale carcass had to be cut up and hauled aboard the mother ship for processing.

One suitable species was the largest toothed whale, the sperm whale which was the role model for Herman Melville's *Moby Dick.* The sperm whale was chosen partly because it spends considerable time swimming slowly on the surface between dives, and also because its carcass floats and produces very high quality oil.

Otherwise, the main target species were the Northern Right Whale, Bowhead Whale and Grey Whale in the northern hemisphere and the Southern Right Whale in the southern hemisphere. All of these whales are large, ranging from 30 to 100 tonnes as adults, and contain lots of valuable oil which keeps their carcasses afloat. They also have huge heads occupying up to one third of the total body length and containing particularly large baleen plates (see Q75), which was another attraction to the whalemen. But their principal attraction was their feeding behaviour. Right whales, and the Bowhead, all feed mainly at the surface by swimming slowly along with their mouths wide open and filtering small food organisms. This made them vulnerable to harpooners in rowing boats, whereas the much faster swimming rorquals (see Q74) were not exploited until steam-propelled whale catchers and explosive harpoon guns were developed in the second half of the nineteenth century.

74 What are rorqual whales?

The rorquals consist of six whalebone whale species, ranging in size from the Minke whale at about 10 metres long and weighing up to 30 tonnes, to the largest of all whales, the Blue whale, reaching a length of 33 metres or more and a weight of up to 190 tonnes.

They are all placed by zoologists into the family *Balaenopteridae*. The common name rorqual, derived from the Norwegian word *rorhval* meaning furrowed or grooved whale, refers to the many skin folds or throat grooves extending from beneath the lower jaw to behind the flippers in all the species. These grooves open out during feeding to increase greatly the volume of the mouth and rorquals feed by taking in a large volume of water, closing the mouth and forcing the water through the baleen plates and out of the sides of the mouth by raising the tongue and re-folding the grooves. In contrast to the right whales that feed by swimming rather slowly, open-mouthed, and filter small food particles on their baleen plates, the rorquals tend to be more active and some species, including the Minke and Humpback, feed on faster swimming and larger prey like shoaling fish.

75 What is whalebone or baleen?

Despite its name, whalebone has nothing to do with bone. It is actually made from keratin, very similar to the stuff that makes up horn, hair and our finger and toe nails. It forms a series of plates edged with bristles, hanging from the upper jaws of whalebone whales (see Q62), used to sieve small food organisms out of the water. The plates are also called baleen, from the old French word *baleine*[33] meaning whale, and the whales that have these plates are consequently also called baleen whales. So the term 'baleen whale' is tautological, literally meaning 'whale whale'.

The size, shape and arrangement of the plates vary in different species depending upon their particular diets. In the Bowhead whales and Right whales (see Q73) the head is huge and they feed by skimming the surface with their mouths open. The 200–400 baleen plates on either side of the jaw are long and slender in these species, 2–3 metres long in the Right whales and 4 metres or more in the Bowhead.

In contrast, the rorquals (see Q74) frequently feed on somewhat larger food,

---

33   The word baleine is still used in French for all big whales with baleen plates, but the plates themselves in French are called *fanons*, a word that applies to more or less any pendulous fold of skin, like a cockerel's wattle, a horse's fetlock or a cow's dewlap.

including fish, often using a gulping behaviour in which a large volume of water is taken into the mouth and forced out of the sides of the mouth through the baleen plates. Their baleen plates tend to be rather shorter and broader than in the Right whales and may have coarser bristles. However, the largest whale of all, the Blue whale, has 250-400 plates on each side of the jaw, each being up to 90 cm long, 50 cm wide and with the finest bristles among the rorquals, in line with its diet predominantly consisting of krill (see Q47).

Because of its strength, combined with flexibility, whalebone was greatly sought after in the days before the availability of plastics. Sadly, it was one of the reasons whales were hunted. It is often said that whalebone was used for stays in ladies' corsets, emphasising the triviality behind the demise of such regal creatures. But although it was certainly used for this purpose, whalebone was also used for more mundane uses like carriage drivers' whips and umbrella ribs.

76 How do whales sing?

The LP *Songs of the Humpback Whale,* initially issued in 1970 by CRM Records and later by Capitol Records, brought to the notice of the public the huge range of eerie and plaintive songs of these huge leviathans. But the Cetacea, the group that includes all the whales, dolphins and porpoises, produce between them a vast variety of sounds produced by different techniques and for different purposes.

The toothed whales (the Odontocetes, see Q62) do not produce the long, low frequency sounds that are widely known as whale song. Instead, they produce rapid bursts of high frequency clicks and whistles, using a mechanism that is not too dissimilar from the one we use. Toothed whales have a narrow section of the nasal passage linking the blowhole with the lungs, which is analogous to our larynx but is rather different from it. Surrounding this narrow passage are muscular phonic lips that can be contracted or expanded at will. As air passes through them, the lips vibrate and these vibrations are eventually transferred to the melon, a fat-filled lump in the whale's forehead region that is believed to concentrate the resulting sound into narrow beams used, at least partly, for echo-location, much as bats do. And the air can apparently be used more than once so, instead of being expelled from the blowhole, it can be recycled and passed through the phonic lips for a second time. The returning echoes are apparently channelled back to the ear through sinuses in the lower jaw.

In contrast, the Mysticetes, or whalebone whales, do not have phonic lips. Instead, they have a larynx much more similar to ours. But unlike ours, the Mysticete's larynx doesn't contain vocal chords, so the precise mechanism of sound production remains a bit of a mystery. However, at least as intriguing is why these leviathans make such emotive and outlandish sounds. See Q77 for an attempt to provide an answer.

77 Why do whales sing?

Q76 pointed out that whales and dolphins produce a wide range of sounds for several different purposes. But why are they such a noisy lot? Well, basically because sound travels through water much more efficiently than light, and about five times as fast as it travels through air.

It is well known that toothed whales, including several dolphin species, use sound for echo-location, that is to 'see' their environment without using vision. This is particularly important for those living in muddy rivers, such as the Pink Dolphin or Boto in the Amazon and the Yangtze river dolphin or Baiji. But lots of experiments have been conducted with captive open ocean dolphins, demonstrating their ability to discriminate between similar objects using sound alone.

On the other hand, although the whalebone whales may use sound to detect the sea floor or the presence of large objects in their vicinity, this has not been convincingly demonstrated. Instead, the plaintive moans, snores and groans of the big whales seem to be primarily for communication between members of the same species or group and, at least in the Humpback whale, are used to find a mate since they are particularly vocal during the mating season.

Finally, it seems possible that whale populations can keep in touch with one another over distances of thousands of km by making use of a fairly narrow depth band in the oceans through which sound travels relatively slowly, but very efficiently. This band, called the SOFAR channel (standing for Sound Fixing and Ranging), is also used for communications between submarines.

Ships And Sailors

78 What is the difference between a ship and a boat?

Basically, ships are big and boats are small; but the boundary between these two categories is very hazy.

In the days of sail, things were much simpler. Different types of vessel were given names based on their rig, that is the number of masts they had and the types of sail they carried. Under this system a ship was defined as a vessel with three (or more) masts, each square rigged, that is carrying rectangular sails supported by spars crossing the masts at right-angles. But with the development of steam propulsion this distinction lost its relevance and the name 'ship' became used for more or less any sort of vessel – as long as it was reasonably big.

In contrast, the term 'boat' has always been used for any small craft ranging from tiny rowing boats, through sailing dinghies and yachts up to fishing vessels that were traditionally fairly small, certainly no more than 100 feet, say 30 metres, long and with crews of less than about ten men. Bigger vessels, either wind or engine driven, were all called ships except that, for some reason, submarines, no matter how big, have always been referred to as 'boats'.

With the increasing size of fishing vessels these distinctions became much more blurred and there are now many huge trawlers carrying large crews capable of catching and processing hundreds of tons of fish without returning to port. The term 'boat' for such vessels seems totally inappropriate. Similarly, the cruise industry exclusively uses ships, not boats. From the very smallest, carrying just a few dozen passengers, up to the leviathans with a combined complement of passengers and crew exceeding 5,000, they are undoubtedly ships!

79 What is a fathom?

A fathom is a nautical unit of length equal to 6 feet or 1.8256 metres. The name comes from an old English word, *faedm,* meaning to embrace. It therefore became the name for the distance across the outstretched arms of an average man, that is about 6 feet. Since sailors used the arm-span technique to measure ropes, including sounding lines to measure the depth of the sea, the fathom became the standard unit of depth. Until relatively recently, most maritime countries continued to use the fathom in navigational charts, but it is now rapidly being replaced by the metre. Curiously, the USA has never fully embraced the fathom (or the metre) and American mariners, including marine scientists, still frequently refer to ocean depths in feet.

80 What is a nautical mile?

A nautical mile is the unit used universally by mariners to measure distances at sea. It is defined as the distance subtended at the earth's surface by one minute of arc of latitude. This simply means that if you imagine the earth cut in half through the north and south poles, two lines on the cut surface from the centre to the outside diverging by one minute, or $1/60^{th}$ of a degree, would intersect the surface exactly one nautical mile apart. In fact, this would be true only if the earth was a perfect sphere, which it isn't! Instead, the earth is actually slightly flattened from pole to pole so that the distance subtended by a minute of arc varies a bit with latitude, ranging from about 1,843 metres (6,046.6 feet) at the equator to 1,861 metres (6,105.6 feet) at the poles. The mean of these, 1,852 metres (equal to 6,076.1 feet) is the internationally agreed compromise and is equivalent to 1.15 (or about one and one seventh) English or Statute miles.

81 What are lines of latitude and longitude?

Some readers will no doubt feel that this is far too simple a question to justify its presence in a book such as this. However, I don't apologise for it because I have met quite a few people who find the concepts difficult to grasp and particularly the consequences of the differences between the two in the production of maps and charts and in navigation; so here goes.

The first point to make is that latitude, and particularly longitude, are entirely human inventions and have no independent validity. They are simply imaginary lines on the surface of the earth used to produce a grid of reference points with which to label specific localities and, of course, to enable us to navigate between them. Let's deal with latitude first (see Fig 15).

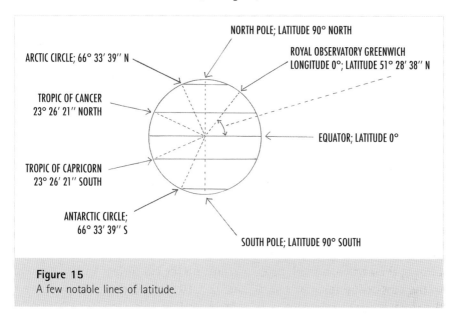

Figure 15
A few notable lines of latitude.

Lines of latitude run east-west parallel to the equator and identify positions on earth north or south of the equator. The equator is a line of latitude, but a rather special one. Since it girdles the centre of the earth at right angles to the planet's axis of spin between the North and South Poles, it divides the earth into two equal halves and consequently has some claim to be recognised as having a unique role. It is accordingly given the somewhat arbitrary label as 0° of latitude. North and south of the equator, the lines of latitude are labelled

---

34 I have in my possession a document that certifies that I crossed the international date line at a latitude of about 180° and a longitude of some 3°. I shall not reveal which vessel my wife and I were on at the time, but it prompted the following piece of doggerel (with the vessel's name replaced by **) which was submitted for publication in the ship's daily newspaper.
'** messed up badly,
** got it wrong.
In writing out the time line certs they mixed up lats and long.
Mine's signed by the captain.
I expect that yours is too,
So instead of finding Tonga we could dock in Timbuktu.'
Perhaps not surprisingly, my contribution was not accepted!

with the angles that they make with the centre of the earth, from 0° at the equator to 90° at the poles. So latitudes are 0-90° north or 0-90° south and it is impossible, therefore, to have a latitude of more than 90°[34]. Each degree is divided into 60 minutes (indicated by a single tick) and each minute is divided into 60 seconds (indicated by two ticks). A precise latitude includes all three, so the latitude of the Royal Observatory at Greenwich, for example, is 51°, 28 minutes and 38 seconds north, written 51° 28' 38" N[35]. Other important latitudes include the Tropics of Cancer and Capricorn at 23° 26' 21" North and South respectively, and the Arctic and Antarctic Circles at about 66° 33' 39" North and South respectively[36].

An important feature of the fact that latitudes are parallel is that the distance between them is always (or almost always) the same (see Q80). Consequently, one second of latitude (that is one sixtieth of a degree of latitude) is equal to one nautical mile or 1.852 km. As a result, latitudes on nautical charts are used by mariners to calculate distances between points on the chart. This is not possible for longitudes for reasons that are explained below.

Lines of longitude run north-south and register the position of localities on the earth east or west of the prime meridian (see below). In contrast to lines of latitude, lines of longitude are emphatically *not* parallel. Instead, they are a bit like the gaps that separate the segments of an orange. As in the orange segments, lines of longitude are far apart at the equator and converge (and indeed cross) at the poles (in the orange these are the point of attachment to the tree and the little dot on the opposite side). Like lines of latitude, lines of longitude are labelled according to the angle they make with the centre of the earth, the full circle around the earth being divided into 360°. Also as with latitude, each degree is divided into 60 minutes and each minute into 60 seconds. But instead of dividing the 360° into four equal 90 degree quarters as is done for latitude, longitudes are simply divided into two halves, one from zero° to 180° west, and the other from zero° to 180° east. So 180° east and 180° west are the same longitude. But what determines exactly where are they?

---

35   Since one second of latitude equals 1/60th of a nautical mile, a latitude given to the nearest second should identify a locality to within about 100 feet or 31 metres. For even higher precision, both minutes of latitude and longitude can be expressed in decimal notation, so that the Royal Observatory's latitude would become 51° 28.633 N.

36   The positions of the Tropics of Cancer and Capricorn, and of the Actic and Antarctic Circles are actually not fixed but depend on the earth's angle of tilt relative to the plane of its orbit round the sun. This oscillates by about 2° with a period of 40,000 years or so. This minor wobble is currently causing the Arctic and Antarctic circles to move nearer to their respective poles by about 15 metres each year, while the Tropics of Cancer and Capricorn are moving by a similar amount towards the equator.

Well, they are opposite the Prime Meridian, that is the line of longitude that carries the label 0° or 0 degrees. Since 1884 this has been a line passing through the Royal Observatory at Greenwich and, even more precisely, through the cross hairs on the eyepiece of the Transit Circle Telescope which had been built at Greenwich in 1850! Since longitude has no useful universally significant reference point like the equator is for latitude, the reference line for longitude is entirely arbitrary. Consequently, over the centuries different map makers had used a variety of different prime meridians ranging from Jerusalem in the east to Washington in the west. However, by 1884, when an International Meridian Conference was held in Washington, D.C. in an attempt to standardise matters, there were only two serious contenders, Greenwich and Paris. When the representatives of the 25 participating nations voted on the proposal to make Greenwich the prime meridian it was passed 22 to 1, with San Domingo (now the Dominican Republic) voting against and Brazil and France abstaining. France did not finally adopt the Greenwich meridian until 1911, but since that time virtually all published maps and charts have used longitudes based on Greenwich as the prime meridian.

82 What is the difference between true north and magnetic north?

True north is the direction towards the geographical North Pole from any point on earth, that is it runs straight up a line of longitude (see Q81) to the pole. In contrast, magnetic north is the direction a compass needle points to in the northern hemisphere. They are usually not the same because, whereas the north geographical pole on a human time scale stays in the same place more or less permanently, the north magnetic pole changes position somewhat irregularly but is rarely, if ever, directly beneath the geographical north pole.

The position of the north magnetic pole was first established in 1831 by James Clark Ross, a Royal naval captain who became much more famous later as an Antarctic explorer. At that time the pole was beneath the wilds of northern Canada, close to the Boothia Peninsula[37], at a latitude of only about 70°N, that is some 1,200 nautical miles south of the geographic North Pole (Fig 16). It moved very little in the next 70 years and was only a few miles away when the famous Norwegian explorer Roald Amundsen located it in 1903. However, during the 20th

---

37   Named, incidentally, by James Clark Ross's uncle, Capt. John Ross, for his sponsor, Felix Booth, the manufacturer of Booth's gin.

century it moved a further 590 nautical miles (1,100 km), slowly at first but, in recent years, at more than 21 nautical miles (40 km) a year, so that it is now thought to be at 83+°N and about 115°W. Similarly, the position of the south magnetic pole is not fixed, but wanders around the south geographical pole just as its northern counterpart does. Its position in 2005 was estimated at 64.53°S, 137.86°E off the coast of Wilkes Land, Antarctica. This locality is actually *outside the Antarctic Circle*, and the magnetic South Pole is moving north-westward at some 10 to 15 km a year. If this were not confusing enough, the imaginary bar magnet passing through the earth and joining the north and south magnetic poles doesn't go exactly through the earth's centre but misses it by more than 500 km.

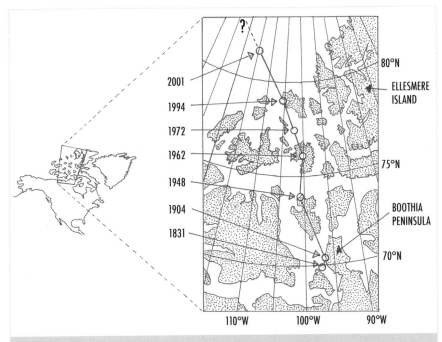

Figure 16
Wanderings of the north magnetic pole in the Canadian Arctic between 1831 and the present day.

The net result of all this is that magnetic compasses do not in general point to the North Pole (or, for that matter, the South Pole), but instead align them-selves with the local magnetic field at an angle to west or east of true north or south. This angle is known as the local magnetic variation[38] and will obviously

38   Also, somewhat confusingly, called magnetic declination, but I'll stick with variation.

change with time as the poles move. To enable map users to make allowance for magnetic variation, most good navigational maps show the geographical north alongside magnetic north at the time the map was published, together with an indication of the direction and rate of change of variation at that location. Of course, there will always be some places on earth where the variation is zero and true north and magnetic north coincide. Lines joining such points are known as *agonic lines*.

Clearly, knowledge of magnetic variation has been crucial for navigators relying upon magnetic compasses. With the development of gyrocompasses (see Q83) and, particularly, satellite navigation, the role of magnetic compasses and the knowledge needed to use them perhaps seems a little redundant. However, both of these modern techniques require electric power to operate them and there have been a number of recent incidents in which quite large and sophisticated vessels have lost all power and have had to rely on the traditional 'string and sealing wax' navigation. Consequently, modern ships, including cruise ships, carry back-up magnetic compasses, and ship's officers are still trained in how to use them.

83 How does a gyrocompass work?

A gyrocompass points to the earth's north geographical pole and not to the magnetic pole. Consequently, its orientation does not vary in space or time as a magnetic compass does (see Q82), nor is it affected by any metal surrounding it.

The heart of a gyrocompass is a gyroscope, a rapidly spinning heavy flywheel, mounted in gimbals, that is a framework that allows the flywheel's axle to move in both a horizontal and vertical axis. This is exactly the same principle used in a child's gyro spinning top. Because of a strange physical phenomenon called the conservation of angular momentum, if the gyro's flywheel is spinning fast enough it will resist all forces tending to change its original orientation.

Both the child's hand-driven top and a pukka electrically powered gyroscope can be set up to point in any direction – and stay there until friction in the top or a power failure in the gyroscope slows the flywheel down. But if this direction is in any orientation other than parallel to the earth's own axis of spin, that is pointing directly to the north and south geographic poles, funny things happen.

The earth rotates on its own axis once every 24 hours. But this spin doesn't affect the gyroscope which 'tries' to point in the same direction irrespective of what the earth is doing. Consequently, to an observer spinning with the earth (as we all are) the gyroscope will itself appear to rotate once every 24 hours. This is clearly no good as a reference for navigation, so the clever trick in a gyrocompass is to set the gyroscope up pointing exactly north-south and then to make use of the earth's rotation to prevent the gyro from wandering away from this orientation.

The usual way this is achieved is to couple the gyroscope with a heavy weight acting as a pendulum. Under the force of gravity the pendulum will hold the gyroscope's axis horizontal with respect to the earth's surface. Except at the equator, this means that the gyroscope's axis will never be exactly aligned parallel with the earth's axis but must continually realign itself as the earth spins. As it does so, any movement away from the north-south orientation will be resisted by the earth's own spinning motion which will cause the compass to precess, that is move back to the north-south alignment where these forces are at a minimum. So once the gyroscope is set spinning and aligned correctly it is a self correcting system and should stay in alignment as long as the flywheel continues to spin. If the power fails and the motor stops, of course, the gyrocompass becomes a pile of pretty useless metal and the dear old magnetic compass, with all its faults, comes back into its own.

84 Why do Greenland and Antarctica look so huge on most world maps?

Partly because they both really are huge, but also, and more significantly, because of the difficulties of representing a spherical world on a flat piece of paper. Let's look at how big they are first.

Antarctica, with a total surface area of about 14.2 million km^2, is almost twice as large as Australia and is therefore the fifth largest continent. Counting Australia as a continent (rather than an island), Greenland, with an area of 2.2 million km^2, is the world's largest island. Being a self-governing territory of Denmark, Greenland also counts as a country, and ranks as the 13th largest country in the world, behind Russia and Canada in first and second spots and China and the USA vying for third place[39]. Nevertheless, despite being right up there with the big boys, Greenland's area is less than one quarter that of Canada. Yet on most maps it looks at least as big as its North American neighbour or even, as on Fig 17, considerably bigger. Fig 17 also shows Antarctica as being almost as big as the rest of the earth's land areas put together.

The reason is that all maps of the earth showing the equator running across the middle distort the shapes and relative sizes of the land masses to a greater or

39 The United Kingdom, incidentally, with an area of 0.24 million km^2, ranks number 79.

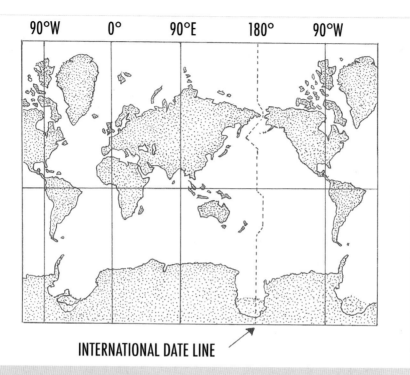

90°W   **0°**   **90°E**   **180°**   **90°W**

INTERNATIONAL DATE LINE

Figure 17
The earth's land masses on a Mercator projection. This also shows (roughly) the International Date Line. The date line is an internationally agreed imaginary line where, by convention, the date changes. Consequently, locations immediately to the left (west) of the date line are always one day ahead of locations to the right (east) of the line. For a good deal of its length the line follows the 180° line of longitude, that is directly opposite the prime meridian at 0° longitude passing through Greenwich. However, so that different parts of individual countries can maintain the same date the line deviates in several places. For example, in the Bering Strait region it deviates to ensure that the whole of the USA and Russia respectively are in the same day, but are on opposite sides of the line, while at about 10°S it deviates to separate Samoa and Tonga. One strange effect of this second deviation is that the two close neighbours keep the same time, but are one day apart, because Samoa is in the western hemisphere while Tonga is in the eastern hemisphere!

lesser extent, and this distortion generally becomes greater the further north or south you go. So the areas near to the poles, like Greenland near the North Pole and Antarctica surrounding the South Pole, are the most distorted of all. The problem, of course, is that the map makers plot the positions of features on the earth's surface on their maps by using a grid usually made up of lines of latitude and longitude (see Q81). Since the lines of latitude girdling the earth are all parallel and equidistant from one another they can be laid out easily on a flat surface.

However, lines of longitude are not parallel and, very inconveniently for map makers, all intersect at the two poles.

Over the centuries there have been dozens of attempts to solve this problem by developing cartographic 'projections', that is techniques for depicting the surface of a globe onto a flat surface. The end results include all sorts of strange shapes, from stars and distorted circles to so-called 'interrupted' maps that look a bit like a flattened discarded orange peel! Many of these have been devised for particular purposes, for example to try to preserve one or other feature such as land shape, area or spatial relationship while sacrificing others. All, to some extent, represent a compromise, but the one with which many of us are probably most familiar distorts high latitude dimensions particularly badly; and this is the one tradition-ally beloved of seafarers for reasons that will become apparent.

When, in 1569, the Flemish cartographer Gerardus Mercator issued his map of the earth based on the projection which came to bear his name[40], it was unashamedly aimed at navigators. Indeed, his Latin title for the map called it a *'New and improved description of the earth corrected for the use of navigation'*. This was because it allowed navigators, for the first time, to lay out on the map a straight line course between two points with a constant compass bearing. This is, after all, the easiest course to steer between two localities (though not neces-sarily the shortest route[41]). Such a course, called a rhumb line, crosses lines of longitude at a constant angle. Because lines of longitude converge towards the poles, all rhumb lines, other than those heading due north, south, east or west, spiral around the earth and eventually reach one or other of the poles. So if you try to lay out a rhumb line on a map based on curving lines of latitude it will follow a curve and not a straight line. Conversely, if you draw a straight line between two points on such a map it will not represent a constant compass heading.

Mercator solved this conundrum at a stroke by the simple, but revolutionary, expedient of drawing his lines of longitude parallel instead of converging and adjusting the spacings between the lines of latitude until the notional rhumb lines became straight! The result was amazing, and it made navigating around much of the earth much easier than it had been previously (though an efficient means of determining longitude was still two centuries away). However, Mercator's projec-tion had big problems at high latitudes and was virtually useless beyond about 70° north and 70° south.

---

40 Not his real name, of course. His family name was Kremer, an old German word meaning merchant or shopkeeper. So when he wanted to Latinize his name he took the Latin word, Mercator, which also means merchant.

41 The shortest route between two points on the earth's surface is a so-called 'great circle', but an explanation of this will have to wait!

This was because, on Mercator's map, in order to make the lines of longitude parallel, the two poles could no longer be shown as single points (which they are, of course), but instead were artificially stretched out all the way across the top and bottom of his world map.

So to go back to the original question, Antarctica looks huge on a Mercator projection because it totally covers the lower margin, which in reality is a single point. At the same time, since Greenland is the nearest major land mass to the north pole, its shape is distorted by Mercator's projection so that it not only looks much bigger relative to more southerly lands than it should, but it is also stretched sideways more in its northerly parts than in the south and therefore looks too top heavy.

85 Why is a knot called a knot?

A knot, equal to one nautical mile per hour, is a unit of speed used by mariners to indicate the velocity of ships, currents and the wind. The name 'knot' refers to the original way in which ships' speeds were measured. This involved using a piece of wood, called a log (presumably because it was originally just that!), attached to the end of a line with knots tied at regular intervals of 30 yards or 90 feet along it. To calculate a ship's speed, the log would be thrown over the side and the line allowed to run out freely as the vessel sailed away from it. The number of knots in the string that passed over the side during one minute, originally timed with an hour glass (actually a 28 second glass) and later with a watch, was the speed of the vessel in 'knots', that is nautical miles per hour.

86 Why port and starboard?

Facing forward, the left-hand side of a ship is referred to as port and the right-hand side as starboard. Starboard is generally considered to be a corruption of steer-board, referring to the steering board or large oar that was hung off the aft end of this side of vessels before the introduction of the modern rudder. To avoid fouling the steering board, the ship would normally be tied up with the opposite or left-hand side against a jetty or quay. This side came to be known as the larboard side, a term probably derived from the old English word ladeboard since this was the loading or lading side of the ship. Because of the danger of confusing the terms larboard and starboard, particularly as steering orders to the helmsman, the term larboard was gradually replaced by port during the early nineteenth century, the new term eventually becoming officially adopted in 1844.

It is said that the word posh was derived from the requirement of wealthy passengers on liners between the UK and India who, to ensure accommodation on the northern, more shady, side of the ship would demand P(ort) O(ut) S(tarboard) H(ome)!

87 What are all those marks on a ship's hull for?

Ships' hulls carry a wide variety of marks which can be a bit baffling if you don't know what you are looking at. Here are a few of the most common ones.

At the front or the back of the ship, and sometimes at both positions, there will be a series of numbers indicating the amount of the ship's hull that is below the waterline. This scale in most ships these days is marked in metres, with the metre marks being clearly labelled 1M, 2M, 3M and so on, and the intervening spaces marked at 10 or 20 cm intervals. However, it is still possible to see ships with these draught marks in feet.

Close to the bows there may be a strange looking mark that consists of a sloping, near vertical line, roughly parallel to the bow itself, joining a curved C-shaped line with the bulge towards the bow. So the 'C' looks the right way round on the port or left-hand side of the hull, and appears backwards on the other side. And this is the clue to its meaning, for it simply indicates that this vessel has a bulbous bow (see Q89) sticking out ahead of the ship below the waterline. Since such 'bulbs' can protrude 3 or 4 metres on some big vessels, this warning is very important to small boats like tugs, pilot launches and fuel barges that have to manoeuvre close to the ship's hull.

Also reasonably close to the bow you may see a circular symbol with a propeller inside it. This marks the position below the waterline of the vessel's bow thruster tunnel. It warns nearby vessels that, during manoeuvring, water may be sucked into or blown out of this tunnel. Very large or specialised vessels may also have similar thrusters near the stern and will, accordingly, carry a similar mark in this position.

Almost anywhere along the hull you may see an area marked as not to be used by tugs. At their simplest these may consist of a fairly obvious statement like 'Do not push', but there are also a variety of alternative markings, most of which are reasonably self explanatory.

Finally, the most familiar, and possibly the most important mark on the side of any merchant ship is the Load Line. This line is commonly referred to as the Plimsoll Line because its use was originally introduced in Britain in 1876 as a result of the campaign by the British Liberal MP, Samuel Plimsoll, to improve the safety of British merchant ships, particularly by stopping ship owners from overloading the vessels. The practice was eventually adopted internationally in 1930 and since then has gone through a series of refinements and is now administered by the International Maritime Organisation (IMO). The current convention governing load lines dates from 1966 but has been frequently updated since that time.

The purpose of the load line is to mark on the side of a vessel the lowest certified safe level, under various circumstances, to which its hull can be allowed to sink into the water.

Any ship floats because its hull beneath the surface displaces a volume of water equal in weight to the total weight of the vessel and its contents. Consequently, the level of the waterline at any time depends upon the vessel's total weight (i.e. including cargo and/or passengers) and the density of the water it is floating in. Since seawater, containing all its salt, is heavier than fresh water, and since warm water (for example in the tropics) is lighter than cold water, the exact level to which a vessel of constant weight will sink before it floats, so to speak, will depend on where it is, and when. The load line symbols can therefore include lots of information to reflect this range of possibilities.

The primary load line, and the one from which all the others are derived, is the Summer Load Line. This line is always shown across the middle of a circle 30 cm in diameter. This line will also be accompanied by a pair of letters that signify which of a number of licensed classification societies around the world determined the line's position for this particular ship and issued its certificate. The letters could be LR for Lloyd's Register, AB for the American Bureau of Shipping, GL for Germanischer Lloyd, BV for Bureau Veritas (Spain), or NV for Det Norske Veritas, IR for the Indian Register of Shipping, and RI for Registro Italiano Navale.

Depending on the nature of the vessel, near the primary line there may be a whole series of short horizontal lines with letters beside them indicating the maximum load line under various conditions. These marks may include any of the following, in descending order down the hull:

TF	Tropical fresh water
F	Fresh water
T	Tropical seawater
S	Summer Temperate Seawater (the same level as the Summer Load Line)
W	Winter Temperate Seawater
WNA	Winter North Atlantic

Finally, some cargo vessels that meet certain special requirements of the certifying agencies may be assigned what are called 'Timber Freeboards'. These are all preceded by the letter L.

88 What does a ship's tonnage mean?

You will often see references to a ship's tonnage, usually as a means of comparing its size with that of other ships. On the face of it, the word tonnage sounds as if it should refer to the vessel's weight. It does not.

The word originally referred to the tax paid on tuns (that is, casks) of wine, but it eventually became a general measure of the cargo-carrying capacity of ships. It is now used as a measure of capacity on which to base a vessel's registration fees, harbour dues, safety and manning requirements and so on. For this purpose two tonnage measurements are mainly used, Gross Registered Tonnage (GRT), and, since 1994, Gross Tonnage (GT).

GRT is a figure that attempts to represent the total internal volume of a vessel (in units of 100 cubic feet or $2.83m^3$) with the exception of non-productive spaces such as crew quarters.

GT, on the other hand, is a figure based on the volume of all of the ship's enclosed spaces and is considered a unitless entity despite being based on volume.

The calculations to derive either of these tonnage figures are complex, but the important thing to remember is that they refer entirely to notional carrying capacity and emphatically not to weight.

However, you may see a figure that really does refer to weight; this is Displacement. It is often called displacement tonnage, but is in fact not a tonnage at all. It is a measure of the actual weight of a ship as if you could hang it on a gigantic spring balance. It is calculated by determining the total volume of the ship below the waterline, that is, the volume of water 'displaced', and then calculating the weight of this water using its density, which will vary, of course, depending on its salt content and temperature.

As an example of how different tonnage and displacement can be, Cunard's *Queen Mary 2* has a displacement of 76,000 tons, but a Gross Tonnage of 148,528.

89 Why do many ships have bulbous bows?

A prominent bulb protruding from the bows below the waterline is a standard feature of most large, modern ships, especially those with long, fairly narrow hulls such as freighters, naval vessels and many passenger ships. Their function is to modify the way the water flows past the hull, reducing the drag and therefore improving the vessel's efficiency.

Although bulbous bows have been widely employed since the 1950s, there seems to be considerable disagreement about exactly how they work, though most accounts suggest that a major part of their effectiveness results from negating the normal bow wave that the ship would otherwise produce. As a normal ship is driven through the water, it produces a wave with a crest immediately in front of the bow and one or more troughs along the length of the hull. This produces variations in the pressure distribution on the hull and increases the resistance to its passage through the water. The bulbous bow produces a second wave, somewhat ahead of the normal bow wave, and the two waves at least partly cancel each other out. This helps to 'flatten' the waterline past the ship, reducing the pressure variations on the hull and therefore the resistance.

Other explanations suggest that water flowing around a bulbous bow improves a vessel's stability, particularly in rough weather, by damping the ship's tendency to pitch, again reducing the resistance.

Whatever the explanation, ships with bulbous bows are thought to have a 12% to 15% greater fuel efficiency than similar vessels lacking them.

Bulbous bows are most effective over rather narrow ranges of speed. Outside this range they can actually decrease a vessel's efficiency. They are also not very efficient on vessels smaller than about 20 m (60 feet) in length. Consequently, they are uncommon on small recreational vessels or those that routinely operate at very variable speeds.

90 How far can you see from a ship at sea? (The simple version for most of us)

Because the earth is spherical, the distance you can see to the horizon depends on how high you are; you can obviously see much further from the top of a mountain than when standing on a beach. Calculating the distance to the horizon on land is complicated by the presence of hills and valleys, buildings and trees and so on. In the words of the old cockney song, 'With a ladder and some glasses you could see to 'ackney Marshes if it wasn't for the 'ouses in between'! But at sea, out of sight of land, the situation is much simpler.

Let's assume that it is a perfectly clear day, that the sea is calm, that you have good eyesight and that the earth is a perfect sphere. *(This last bit isn't quite true, but most of us can ignore it and leave it to the nerd's version.)* Now, if we know how far our eyes are above the surface of the sea, we can work out the distance to the horizon. The proper way to do this is to use dear old Pythagoras's theorem which, you will remember, says that in a right-angled triangle the square on the hypotenuse (the side opposite the right-angle) is equal to the sum of the squares on the other two sides. But at this stage we will cheat a bit with two simple formulae that give the right answer as near as damn it.

First, if you are working in miles the formula is as follows:

$$d = \sqrt{(1.5 \times h)}$$

This means that the distance to the horizon (d) (in statute miles) is equal to the square root of one-and-a-half times the height above the surface (h) of your eyes in feet. To work a simple example, if you are standing in a rowing boat with your eyes 6 feet above the surface, the distance to the horizon is $\sqrt{(1.5 \times 6)}$ (that is $\sqrt{9}$), or 3 miles. And if you are standing on the deck of a ship 30 feet above the sea surface you can see $\sqrt{(1.5 \times 30)}$ (that is $\sqrt{45}$) or 6.7 miles to the horizon.

If you prefer to work in km, the formula is a bit more complicated:

$$D = \sqrt{(13 \times H)}$$

Where the distance to the horizon (D) in km is the square root of 13 times the height of your eyes (H) in metres. To work a couple of simple examples using this formula, if your eyes are 2 metres above the surface the distance to the horizon is $\sqrt{(13 \times 2)}$ (that is $\sqrt{26}$) or 5.1 km. Similarly, if your eyes are 10 metres above the sea, you can see $\sqrt{(13 \times 10)}$ (that is $\sqrt{130}$) or 11.4 km to the horizon.

Easy, isn't it?

91 How far can you see from a ship at sea? (The nerd's version)

Q90 provided a couple of simple formulae to work this out by a bit of cheating. This one is for those of us who want to do it the proper way.

When you are standing on a ship, your line of sight to the horizon forms a right angle at that point with an imaginary line to the centre of the earth (see Fig 18). So the triangle we are solving has the following three sides: one equal to the radius of the earth (let's call it r), a second equal to the distance from our eyes to

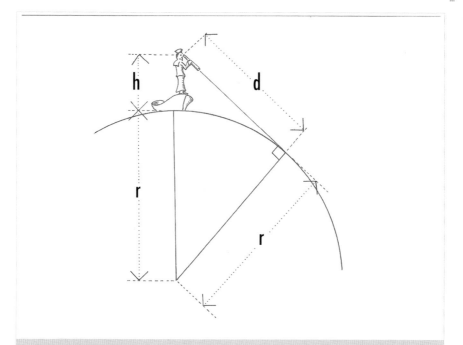

Figure 18
Sight distance to the horizon from a ship at sea. The line of sight (distance d) makes a right angle with an imaginary line from the centre of the earth to the horizon (length r). The right-angled triangle is completed by a line from the earth's centre to the eye of the observer (length r + h). Remember that r, the radius of the earth, is a big number compared with h. So the triangle is a very long and thin one making a very tiny angle at the centre of the earth.

the horizon (let's call it d), and finally the distance from our eyes to the centre of the earth, which equals r plus our height above the sea which we will call h.

From Pythagoras we know that:

$$r^2 + d^2 = (r+h)^2 = r^2 + 2rh + h^2$$

Therefore:

$$d = \sqrt{(2rh + h^2)}$$

Using a figure of 6,371 km as the average radius of the earth, Table 1 shows the distances to the visible horizon from an observer at various heights above the sea surface ranging from 1 metre (say sitting in a rowing boat) to 20 metres (the height of the bridge on a medium sized cruise ship). You might be surprised to see how short the 'horizon distances' are, only about 3.5 km (a little more than 2 miles) from the rowing boat and about 16 km or 10 miles from the 20 metre high deck.

And you might want to work out some other horizon distances for different heights above the sea. If you do, Table 1 also shows you how the results from the simple formulae compare with the longer calculation.

Finally, let's return to the business of whether the earth is a perfect sphere or not. It isn't. Instead it is slightly flattened from pole to pole. Consequently, the *equatorial radius* (about 6,378 km) is a bit bigger than the *polar radius* (about 6,357 km). (We've used an 'average' figure of 6,371 km for our calculations.) As a result, the curvature of the earth's surface is slightly more pronounced or 'tighter' in a north-south alignment than east-west so that the distance to the visible horizon at the equator is slightly less looking north or south than looking east or west. But these differences are so small that we can ignore them for our purposes.

TABLE 1 Distance to the visible horizon in km, nautical miles (Nm) and statute miles (M) from various heights above the sea surface from 1 to 20 metres (see text for details).

1	2	3	4	5	6
Height (metres)	Distance to visible horizon using $d=\sqrt{rh + h^2}$			Horizon distance (km) using $D=\sqrt{13H}$	Horizon distance (miles) using $d=\sqrt{1.5h}$
	Km	Nm	M		
1	3.57	1.92	2.22	3.6	2.22
2	5.05	2.72	3.14	5.09	3.14
5	7.98	4.3	4.96	8.06	4.96
10	11.29	6.09	7.02	11.4	7.02
15	13.82	7.45	8.59	13.96	8.59
20	15.96	8.61	9.92	16.12	9.92

92 How far off can you see land from a ship at sea?

Qs90 and 91 showed you how to work out the distance to the visible horizon from a ship at sea. Of course, you can use the same formulae to work out the distance to the visible horizon from the land, say from the highest point on an island. For example, using the simple imperial formula in Q90, from the top of a hill 1,000 feet high you could see $\sqrt{(1.5 \times 1,000)}$, that is 38.73, statute miles. To an observer in a ship at sea the tip of this hypothetical island should just become visible at this distance plus his own visible horizon distance. So if the seafarer's eyes are 10 feet above the surface of the sea, the highest point of the island should become visible when he is 42.6 miles away from it (that is $\sqrt{(1.5 \times 10)}$ or 3.87 nautical miles for the distance to his own visible horizon plus the 38.7 miles for the visible horizon from the top of the island.

Before the advent of modern satellite-based position fixing and other electronic navigational aids, this information could be quite crucial for seafarers approaching a coastline with good data on the altitudes of prominent features on the land. Consequently, volumes of navigational data traditionally included 'Distance off' tables providing navigators with a quick means of working out how far away they would be when these features appeared or disappeared on particular courses.

Wind, Waves and Weather

93 How does the wind cause waves?

On the face of it, this seems a pretty daft question. Surely, the answer is that it just blows the water, doesn't it? Well yes, but exactly how does it do this?

In 1774, Benjamin Franklin wrote that *'Air in motion, which is wind, in passing over the smooth surface of the water, may rub, as it were, upon that surface, and raise it into wrinkles, which, if the wind continues, are the elements of future waves'*, and he was right!

What he was saying was that if two fluids in contact are moving at different speeds, then the friction between them will cause a transfer of energy, in this case from the air to the water, mostly producing waves and, to a much less extent, currents.

You can see the beginnings of the wave generating process when you blow across the surface of a cup of tea, producing the tiny wavelets or Benjamin Franklin's 'wrinkles'. If the wind continues, and particularly if it strengthens, the wrinkles become 'the elements of future waves', as Franklin suggested. This seems to be because, once the wind speed exceeds about 1 metre per second (that is 2 knots), the wind presses on the back (upwind side) of the wave to produce a slightly increased pressure, while the front of the wave (the downwind side) is somewhat sheltered from the wind and experiences a slight decrease in pressure. This difference in pressures pushes the wave along in the direction of the winds. But in the open sea, although the wave clearly moves, there is little or no corresponding movement of water. This is obvious if there is something floating on the surface of the water, like a buoy or a stationary small boat.

As the wave passes, the buoy or boat will simply rise and fall, but remain more or less in the same place. Actually, if you watch very carefully you will see that they will move with the wave's movement at the crest and against the wave at the trough, finishing very slightly downwave from where they started. This is because the energy in the wave moves the individual water particles at the surface in an almost circular or orbital vertical path a   ameter of this orbital path is equal to the wave's height, th   is   ince between the trough and the crest. The orbit is almost circular,   because the forward movement at the crest is slightly larger than   ovement at the trough, hence the slight movement with the wave. This   called wave drift, and is extremely small relative to the wind speed.

But if the wind blows in mo ᵕ o   ess the same direction consistently over the ocean surface it will drag the s ace layers with it as a current, typically reaching a speed of about 3% of th wind s ed. But whereas waves travel more or less exactly in the same direction as t'   ind generating them, wind driven currents move at an angle approaching 45   ne right of the wind in the northern hemisphere and to the left in the southern .iemisphere, all because of the rotation of the earth (see Q19).

94 What is the Beaufort wind scale?

The Beaufort scale is a system for describing wind speeds using a series of 13 categories or forces, from Force 0 indicating winds of less than 1 knot (or 1 nautical mile per hour, see Q80), to Force 12 indicating winds of speeds greater than 65 knots and referred to as hurricanes.

The system is named after its 'inventor', Sir Francis Beaufort (1774–1857), a British Rear Admiral who served as Hydrographer[42] from 1829 to 1855. In fact, rather than inventing the scale, Beaufort's genius was to refine wind scales that were already in use, and to define the categories in terms that were easily understood by his fellow mariners. His system was adopted by the British Admiralty in 1838 to be used by all Royal Naval ships and, with subsequent refinements, has become almost (but not quite, see Q95) universally adopted for maritime weather reporting and forecasting.

In Beaufort's day all ships' deck officers would have been very familiar with the effects of wind speeds on the sailing qualities of their vessels. Beaufort's descriptions

---

[42] Hydrographers study all aspects of the oceans relevant to navigators, including wind, waves, currents and tides. But they are also concerned with producing accurate charts of shore lines and dangers such as rocks and shallow banks. Royal Naval officers had been producing such charts since the reign of Henry VIII, but there was no formal organization within the navy to supervise these activities until the Hydrographic Office was established in 1795, with an officer, the Hydrographer, in charge of it. The post would probably have gone to Captain James Cook if he had not been killed by Hawaiian natives at Kealakekua Bay in 1779. As it was, the first Hydrographer was Alexander Dalrymple and Francis Beaufort was the fourth to hold this prestigious post (see Ritchie, G.S. *The Admiralty Chart*, The Pentland Press, 1995).

of the wind force categories were therefore based on details of the sails that could be used in each category rather than the actual speed of the wind, ending with the immortal definition of a force 12, or hurricane, as 'that which no canvas could withstand'. Nowadays, each category is associated with a specific range of wind speeds, for example 34–40 knots in the case of a 'gale force 8'. But the descriptions of the effects of each category (see below) refer only to the effects on the sea surface in deep water, since the effects on sails would be irrelevant for most modern mariners.

Beaufort scale force	Knots	Description[43]	Height of sea (ft)	Deep sea criteria
0	0–1	Calm		Flat calm. Mirror smooth
1	1–3	Light airs	1/4	Small wavelets, no crests.
2	4–6	Light breeze	1/2	Small wavelets, crests glassy but do not break.
3	7–10	Light breeze	2	Large wavelets, crests begin to break.
4	11–16	Moderate breeze	3.5	Small waves, becoming longer, crests break frequently.
5	17–21	Fresh breeze	6	Moderate waves, longer, breaking crests.
6	22–27	Strong breeze	9.5	Large waves forming, crests break more frequently.
7	28–33	Strong wind	13.5	Large waves, streaky foam.
8	34–40	Near gale	18	High waves of increasing length, crests form spindrift.
9	41–47	Strong gale	23	High waves, dense streaks of foam, crests roll over.
10	48–55	Storm	29	Very high waves, long overhanging crests. Surface of sea white with foam.
11	56–65	Violent storm	37	Exceptionally high waves, sea completely covered with foam.
12	Above 65	Hurricane	–	The air filled with spray and visibility seriously affected.

43 These terms vary a bit. For example, forces 8 and 9 are often referred to as gale and severe gale respectively.

95 What does the Shipping Forecast mean?

The shipping forecast transmitted on BBC radio 4 four times each day is one of the most distinctive and familiar set of sounds to which the British public are exposed. Yet to understand what the announcer is saying you have to be something of an aficionado, or an anorak, or both. This is mainly because it is a masterpiece of word economy, being limited to an absolute limit of 370 words to put over a great deal of complicated information to an audience of mariners who may be listening in far from ideal conditions. To do this, it not only employs an absolutely rigid format with which its regular listeners are very familiar, but it also uses a system of 'shorthand', including a number of words which, in general usage, have fairly broad meanings, but in this context are very precise. Let's have a look at the various elements.

After in introductory 'And now for the shipping forecast, issued by the Met Office . . . at xxxx today . . .' there will be a short statement of gale warnings if winds of force 8 or more on the Beaufort scale (see Q94) are forecast. If not, then the announcer will give the *General synopsis,* that is a very brief account of the distribution and expected movements of high, and particularly low pressure areas in the region. Positions are referred to using the system of sea areas (see Q96) with which the professional listeners will be very familiar.

The forecast for each of these areas in turn, or in batches if the forecast warrants it, will now be read out in a fairly strict clockwise order beginning with North and South Utsire (see Q96) off the southern coast of Norway, then proceeding southwards through the North Sea, westwards along the Channel, then northwards into the Irish Sea and around the west of Ireland to end to the north and west of Scotland in Fair Isle, Faeroes and South-east Iceland. This treatment will include forecasts for Biscay, bordering the Atlantic coasts of France and northern Spain, and also for Fitzroy, a huge open ocean area to the west of Biscay. The 0048 broadcast will also include a forecast for sea area Trafalgar, off the coast of Portugal, which is not mentioned in the other forecasts unless there are gales expected.

Each area forecast similarly follows a strict format. The wind direction is given first, together with its strength on the Beaufort scale, emphasized by name for categories of 8 and above, for example Gale 8, Severe Gale 9, Storm 10, Violent Storm 11 and Hurricane force 12, this being the only context in which the word *force* is used in the broadcast.

Next comes a mention of precipitation, if any, and finally the visibility using shorthand 'code' words. Thus 'good' visibility means more than 5 nautical miles (see Q80), 'moderate' means between 2 and 5 nautical miles, 'poor' means between 1,000 metres and 2 nautical miles, and finally 'fog' means less than 1,000 metres.

When severe winter conditions could lead to the accumulation of large quantities of ice on a vessel's rigging and superstructure, which might threaten its stability, then the last item will be a warning of light, moderate or severe icing. This usually applies only to South-east Iceland.

Finally, generations of night hawks and ship's officers will also know that the 0048 broadcast each day is traditionally preceded by the playing of the mellow musical piece *Sailing By* composed in 1963 by Ronald Binge. Binge is probably even better known for his arrangement of *Charmaine* for the Mantovani orchestra in 1951, which received worldwide acclaim.

Happy listening.

96 Where on earth is Utseera?

We've all heard the word, but few of us have a clue what it refers to. It is actually spelled Utsire and is an area (or rather two, because there is a North and a South Utsire) of the northern North Sea off the coast of southern Norway. We know the word because it invariably starts the regional part of the familiar shipping forecast (see Q95) broadcast by BBC radio 4 four times a day beginning at 20 past 5 in the morning, then at 1201, 1754 and finally at 0048 as Radio 4 closes down and hands over to the World Service for the wee small hours.

North and South Utsire are two of a total of 31 sea areas into which the seas around north-western Europe are divided for British shipping weather forecasting purposes (see Fig 19) with emotive names like Viking, German Bight, Fastnet, Rockall and South-east Iceland.

The sea areas have been much the same for the last 60 years, though older listeners will have noticed some significant changes. For example, Fisher was created in 1955 and Heligoland was renamed German Bight the following year. In the early 1980s the sea area Minches off north-western Scotland was merged with Hebrides, while Viking was reduced in size with the introduction of dear old North and South Utsire (already recognised by the Norwegians). Finally, the old Finisterre was renamed FitzRoy in 2002 to honour the founder of the Met Office (and Charles Darwin's captain on HMS *Beagle*) and to avoid confusion with a Spanish area of the same name.

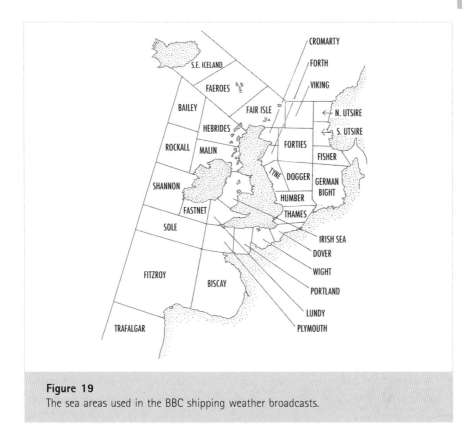

Figure 19
The sea areas used in the BBC shipping weather broadcasts.

97 Why are the Trade Winds called the Trade Winds?

The trade winds are the steady and regular winds blowing in bands on either side of the equator and extending to about 30° north and 30° south. In the northern hemisphere the trades blow from north-east to south-west and are therefore called the north-east trades, while those in the southern hemisphere blow from the south-east to the north-west and are called the south-east trades. They are separated by the doldrums, a region usually close to the equator itself, where the winds are generally very gentle or non-existent.

The trade winds are often said to be so named because in the middle ages they enabled sailing ships from Europe to sail to the west, to the New World, to carry out 'trade', and to return home in the prevailing westerly winds at higher latitudes between about 30° north and 60° north. Although this seems a very reasonable explanation, it is much more likely that the use of the term trade, both in the sense of 'business' and for the winds, arose independently from the same old Low German and Saxon words *trade* and *trada,* which roughly meant track or trail and give us the modern word *tread.* From this origin it is easy to see how the word came to be used for any regularly used track or path, while it is a small step from this meaning to its use for an habitual occupation or business. But according to the Oxford English Dictionary the 'trade wind' combination dates back to the 17th century when it was used in the expression 'blows trade' referring to *any* wind blowing consistently in the same direction. Only later was the term restricted to the current use, so the answer to the original question is that the term trade wind simply means one that blows constantly in the same direction.

98 What causes the Trade Winds?

Like all winds, the trade winds are caused by differences in atmospheric pressure at different points on the earth's surface, with air moving from high pressure areas to low pressure areas until the pressure difference disappears. In general, the strength of the wind is directly related to the pressure difference: the bigger the difference, the stronger the wind.

In turn, the atmospheric pressure at any point on the earth's surface is produced by the total weight of air above it and pressing down on it as a result of gravity. Differences in pressure are caused mainly by differences in air temperature, with warm air being relatively light and producing lower pressures at the earth's surface, while cold air is heavier and results in higher pressure. Lots of factors act together in many areas of the earth to produce variations in the winds and rain over short time scales of hours or even minutes, resulting in the changeable weather so typical of areas such as northern Europe. But in large areas of the tropics and subtropics, where the trade winds occur, the weather situation is generally much more stable. Let's see why.

Fig 20 is a simplification of a very complicated system, but it shows that the surface of the earth is encircled by five great bands or zones of varying pressure, and stability, running roughly parallel to the equator. The key to the trade winds is the central three of these pressure zones, the equatorial low pressure zone running, as its name suggests, more or less along the equator, and two subtropical high pressure zones running respectively, at roughly 30° north and 30° south. Air masses from both subtropical high pressure zones move towards the equatorial low pressure zone as winds that converge to justify the equatorial zone's alterna- tive name of the Intertropical Convergence Zone or ITCZ. If the earth did not rotate, these winds would blow directly south and north, but as a result of the

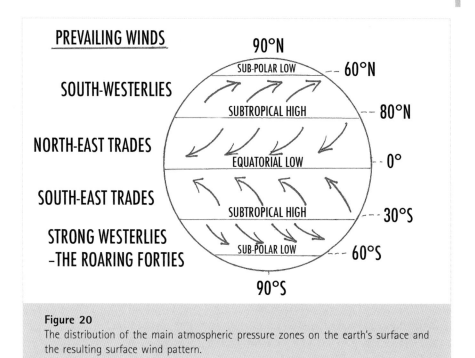

PREVAILING WINDS

90°N

SUB-POLAR LOW --- 60°N

SOUTH-WESTERLIES

SUBTROPICAL HIGH --- 80°N

NORTH-EAST TRADES

EQUATORIAL LOW --- 0°

SOUTH-EAST TRADES

SUBTROPICAL HIGH --- 30°S

STRONG WESTERLIES
-THE ROARING FORTIES

SUB-POLAR LOW --- 60°S

90°S

Figure 20
The distribution of the main atmospheric pressure zones on the earth's surface and the resulting surface wind pattern.

rotation and Coriolis force (see Q19) they actually blow towards the southwest and the north-west, resulting in the north-east and south-east trades.

Outside the subtropical high pressure zones there are two further low pressure zones, the north and south sub-polar lows. Air flows from the two subtropical highs to each of these low pressure zones, producing the prevailing south-westerly winds of the North Atlantic and the powerful westerlies of the roaring forties, fearsome fifties and screaming sixties of the Southern Ocean. But the sub-polar lows, and particularly the northern hemisphere one, are not nearly as constant as the equatorial low and the subtropical highs. Consequently, it is largely variations in the strength and position of the northern sub-polar low pressure region that gives northern Europe, North America and Asia their changing weather patterns.

We can't leave the topic of trade winds without pointing out that earth's near surface winds, like the ocean's near surface currents, form part of a much more complex system. Just as the deep waters of the oceans move around in a layer cake of currents at different speeds and in different directions at different depths, so does the air in the atmosphere at different altitudes. Fig 21 summarizes this system, very simplistically, rather like Fig 3 did for the deep ocean currents, showing that above each of the major areas between the high and low pressure zones there is a vertical circulation of air in which the low level winds, including the trades, are mirrored by winds in the opposite direction at an altitude of 10-12 km. These circulations are called cells, and are named in honour of the scientists who first explained them adequately. Those above the trade winds, that is between the equatorial low and the two subtropical highs, are called Hadley Cells after George Hadley, an 18th century English amateur meteorologist who first proposed this circulation as an explanation of the trade winds. The two cells to the north and south of the trade winds belt are called Ferrel Cells in honour of William Ferrel, a 19th century American meteorologist who refined Hadley's work and produced the first convincing explanation of the mid-latitude circulation.

Looking specifically at the Hadley cells, as the two sets of trade winds approach the equatorial zone they become warmer and pick up more and more moisture from the ocean. Once near the equator the air masses become even warmer and lighter under the sun's influence and begin to rise, reducing the pressure at the surface. As the air rises, it starts to cool so that the contained moisture condenses to form the huge cumulonimbus clouds so typical of the tropics, towering thousands of metres high and dumping their load of water as violent tropical thunder storms. By the time the air reaches the top of the troposphere (effectively the limit of the planet's weather)[44], it has lost virtually all its water. It now moves

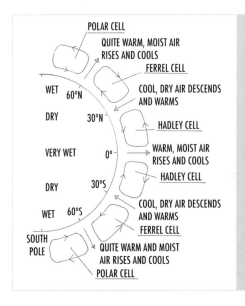

Figure 21
Distribution of atmospheric circulation cells over the surface of the earth. Note that the vertical scales of the cells are hugely exaggerated for clarity. Since the cells are only 10 km or so deep, while the distance between the north and south poles is about 12,700 km, if the cells were drawn at the same scale their upper and lower winds would be lost in the thickness of a single line.

---

44   The troposphere is the lowest part of the earth's atmosphere, as distinct from the next layer, the stratosphere. Most of the phenomena affecting our day to day weather take place in the troposphere. Temperature generally *decreases* with altitude in the troposphere, whereas in the stratosphere temperature *increases* with altitude. Between the two is a layer called the tropopause which effectively acts as a cap on the troposphere. The average height of the tropopause (and therefore the thickness of the troposphere) is about 11 km, but varies from up to about 20 km in the tropics to 7 km at the poles.

towards the poles in both hemispheres but, because of the Coriolis force (see Q19), in both cases the movement is deflected towards the east, that is more or less in the opposite direction to the underlying trade winds. Finally, when these high level air masses reach the subtropical high pressure regions they descend towards the earth's surface, thus completing the Hadley Cell loop. Because the descending air contains little moisture, the subtropical high pressure regions are characterised by very low rainfall – which is why the planet's deserts are concentrated around latitudes 30° north and south.

Hadley Cells are relatively simple, being driven by equatorial solar heating. The polar cells, beyond 60°N and S, are similarly simple, being also driven by convection; even at 60° of latitude in both hemispheres the air masses are sufficiently warm and moist to promote a convective circulation similar to those in the Hadley Cells, but rather less powerful. However, between the Hadley and Polar Cells in each hemisphere there is a Ferrel cell in which the vertical circulation is in the opposite sense, that is with poleward flowing air at ground level and an equatorward flow at high altitude. Ferrel Cells are secondary features, that is they depend for their existence on the cells on either side of them, acting a bit like a passive ball bearing between two rotating rollers. Consequently, Ferrel Cells and their apparently characteristic features, such as the prevailing westerlies, are much less stable than those of the Hadley and Polar Cells, particularly in the northern hemisphere. As a result, they can be disturbed fairly easily, especially by variations in the strength and position of the sub-polar lows, leading to the variable weather so typical of temperate latitudes, as noted above.

99 How deep are the effects of waves felt?

Q93 showed that as waves pass over the sea, water particles at the surface move in more or less circular vertical paths, with a diameter equal to the wave height. In deep water, individual water particles beneath the wave also move in circular paths, but with the diameter becoming smaller and smaller with increasing depth as the wave's energy gets used up, or dissipated. Eventually, at a depth equal to half the wave length (that is the distance between two successive wave crests), water movement due to the passing waves reaches zero. An important consequence is that, since the wave lengths produced by the most severe storms are no more than 300 metres or so, submarines have to submerge only to about 150 metres to avoid any effects. The same is true, of course, for any marine animals.

100 Why do waves break when they reach the shore?

Q93 explained that when waves are moving across the surface of the sea in deep water, the energy in the wave moves the individual water particles in more or less circular vertical paths. At the surface the path followed has a diameter equal to the wave height, that is the vertical distance between the wave's crest and trough.

With increasing depth, the water particles continue to follow circular paths, but the diameter of the circle becomes smaller and smaller until it reaches zero at a depth equal to half the wavelength, that is the distance between two successive crests (see Q99).

If the depth is less than half the wavelength, however, friction between the moving water and the sea floor has a number of effects.

First, the original circular paths of the water particles become squashed into more and more flattened ovals until, very close to the bottom, the water simply moves backwards and forwards, often moving the bottom sediments back and forth as it does so.

But the shallowing water also has much more obvious effects on the waves. In deep water, the speed at which waves travel across the surface of the sea is entirely dependent upon the wave length; the longer the wave length, the faster the wave travels. In shallow water, on the other hand, friction with the sea floor slows the waves down, so that the shallower the water, the slower the waves. As a result, the waves start catching up with one another so that the distance between the crests decreases and the waves themselves become higher and steeper until they eventually break, to dissipate the wave's energy.

Depending on the weather conditions and the steepness of the seabed, the breaking waves take on one of the following four major types.

1. *Spilling breakers.* These are characterized by foam and turbulence at the wave crest, usually starting some distance from the shore. The spilling starts because water at the top of the wave moves faster than the wave itself and is typical of waves approaching a gently sloping shoreline and particularly during a storm.
2. *Plunging breakers.* These are the classic arched surf-rider's waves and are typical of long swells generated far away and approaching gently sloping shores, for example the popular surfing beaches of south-western England and California. The shorter waves produced by local storms may produce plunging breakers on steeper shores.

3. *Collapsing breakers.* These are quite similar to plunging breakers except that the wave front simply collapses into foam rather than curling over. They are typical of moderately steep slopes and moderate wind conditions.
4. *Surging breakers.* These are typical of long, low waves approaching very steep beaches. Instead of breaking, these waves dissipate their energy by simply surging up and down the steep beach.

101 How big are the biggest oceanic waves recorded?

Roughly 30 metres, or a little over 100 feet high, from crest to trough. Measuring or estimating the height of such waves from a ship at sea is extremely difficult unless you have access to rather sophisticated technology. Consequently, there have been many claims by seafarers to have experienced huge waves, sometimes well in excess of 30 metres high, but in the absence of corroborative evidence they cannot be substantiated.

The record for the highest wave ever recorded with reasonable certainty was an observation made from the US Naval vessel *Ramapo*, a 146 metre long oil carrier, during a storm in the northern Pacific on 7 February 1933. The storm lasted for a week, giving the crew ample time to observe the huge seas produced by the 60 knot winds. At one point an officer on watch on the bridge saw the crest of a wave approaching from the stern aligned with the crow's nest on one of the vessel's masts when the ship's stern was in the trough. Subsequent calculation gave an esti- mated wave height of 34 metres or close to 112 feet. At the same time the wavelength, that is the distance between two successive crests, was estimated at 342 metres. With a period of 14.8 seconds between the passage of successive crests, the velocity of the wave was calculated at 23 metres/second or 46 knots.

Although this observation is well documented, it remains an estimate and therefore subject to human error. However, in February 2000 the British research vessel RRS *Discovery* encountered severe westerly gale force winds averaging 21 metres/second (42 knots) in the North Atlantic close to the islet of Rockall and some 250 km west of the Scottish mainland. Because of the severe weather condi- tions the ship was forced to heave to, that is to head into the approaching waves and simply maintain its position and heading. It was fitted with a Shipborne Wave Recorder (SBWR), an instrument first developed by British scientists in the 1950s specifically to record wave heights from ships. The version fitted to *Discovery* had a pair of accelerometers and pressure sensors mounted on either side of the ship below the water line. The information coming from these sensors is integrated to make allowance for the ship's roll and different heights of the water on either side. The net result was that it recorded individual waves up to 29.1 metres (95.5 feet) from crest to trough, and a maximum significant wave height of 18.5 metres (almost 61 feet). The significant wave height is defined as the average height (crest to trough) of the biggest one third of the waves in a series.

This is now recognised as the biggest oceanic wave ever recorded by scientific instruments[45].

102 What are rogue waves?

Rogue waves, also known as freak waves or monster waves, are surface waves that suddenly appear, seemingly out of nowhere, and are significantly bigger than the more typical waves being experienced at that time. They are defined scientifically as waves with a height more than twice the significant wave height. In turn, the significant wave height is defined as the average height of the largest one third of all the waves. So rogue waves are not necessarily the biggest waves encountered at sea (see, for instance, Q101), but they are important because they are surprisingly large waves for a given sea state, and their danger to shipping arises from their unforeseen arrival.

Rogue waves have been a part of maritime culture for centuries and there are many stories of vessels encountering monstrous waves moving across the ocean like a huge wall of water. Such waves have often been cited as the cause of unexplained ship losses though there is little or no direct evidence in support of these explanations. However, the existence of rogue waves is no longer in doubt and research in recent years has demonstrated that they are much more common than most wave theory models would predict. Consequently, there is still considerable discussion among scientists about the causes and how to forecast where and when they are to be expected.

One possibility is that very large waves can develop where strong currents flow against the prevailing wind direction, shortening the wavelength and therefore increasing the height, possibly a bit like the production of tidal bores (see Q111). This idea is based at least partly on a concentration of rogue wave reports in the region of Cape Agulhas off the southern tip of Africa, where such a current and wind regime exists.

---

45   The Rockall region is renowned for its high winds and waves and the previous highest instrument-recorded wave in this area was a 26.3 metre (86 feet) wave measured in December 1972 by the Ocean Weather ship *Weather Reporter*.

Another suggestion is that particular coastline and/or seabed shapes funnel 'normal' surface waves in such a way that they interact, sometimes cancelling one another out, at other times adding together to produce unusual waves much as sound waves can interact to form beats.

Finally, a much more complex explanation suggests that so-called non-linear processes (which I certainly don't understand) cause quite normal, and relatively stable, waves to produce an unstable wave form that basically extracts much of the energy from its neighbours and grows into a monster. However, the theory suggests that the new wave will be unstable and will therefore last only a short time, perhaps a few minutes, before dissipating its energy by breaking or transferring some of its strength back to its neighbours. Such waves would be extremely difficult to forecast, but any ship caught by one would also have to be extremely unlucky. This explanation seems to fit the capsize of the Japanese fishing vessel *Suwa-Maru No 58* in the northern Pacific in June 2008 with the sad loss of 17 of her crew of 20. The subsequent investigation, briefly reported in the *New Scientist* in March 2009, concluded that low and high frequency components of the fairly moderate seas being experienced by the fishing vessel somehow interacted and combined to produce the sudden huge wave that overwhelmed it.

Unlucky or not, there are some classic examples of fairly incontrovertible rogue waves, sometimes involving famous vessels. For example, in 1942 the Cunarder *Queen Mary,* then being used as a troop ship, almost capsized some 600 nautical miles west of Scotland when she was hit broadside by a wave estimated at more than 90 feet (27 metres) high.

Fifty three years later (in 1995) another Cunarder, this time *Queen Elizabeth 2* (finally retired to Dubai in 2008), encountered a similarly sized wave, again in the North Atlantic, which, said her Master, 'looked like the white cliffs of Dover'; a wall indeed!

Finally, one of the few videos of a rogue wave appeared in a 2005 episode of the popular television series *Deadliest Catch,* about the north Pacific crab fishery. In this short video, now available on the internet, the fishing vessel *Aleutian Ballad* is shown battling her way through rough, but not remarkable seas, when she is suddenly hit on the starboard side by a much bigger wave. The vessel is forced over onto her port side, causing the engines to stop because of loss of oil pressure. A second, similar wave would probably have sunk the vessel. But there wasn't one, and the ship and crew survived with only minor injuries.

103 What are tsunamis?

Tsunamis are waves caused when a part of an ocean is suddenly displaced. The reason can be an earthquake, a sudden underwater landslide, itself possibly the result of an earthquake, an impact from an extra-terrestrial body such as an asteroid, or even a man-made underwater explosion. The resulting rapid displacement of water causes a series of waves which move at very high speeds across the ocean surface and can cause devastation when they strike land, as in the case of the disastrous Indian Ocean tsunami of 26 December 2004 which killed over 300,000 people.

Tsunamis are often referred to as tidal waves, but this is a misnomer since they have nothing to do with tides. The Japanese name means 'harbour wave' and refers to the fact that they are normally noticed only in inshore waters such as harbours. This is because the wavelength of a tsunami is huge, up to hundreds of kilometres, but while it is crossing deep water its wave height is very small, often no more than a metre or so. Q93 and 100 explained that, for normal wind-driven waves, the speed of travel across the surface in deep water is entirely dependent upon the wavelength; the longer the wavelength, the faster the wave travels. But once the water shallows to half the wavelength or less, friction with the bottom slows the waves but increases their height. The same thing happens to tsunamis, but because their wavelength is many times the depth of even the deepest oceans their speed is always governed by the water depth. Nevertheless, they can travel at speeds of 800 km/hour so that it can take 20 to 30 minutes for a tsunami to pass a ship at sea from one crest to the next and therefore be extremely difficult to detect.

When the tsunami approaches shallow water, however, the situation changes dramatically. Even in moderately shallow water the waves' speed and wavelength will decrease by a factor of ten or so and, as a result, the wave height will increase significantly to produce a very visible wave. But it is only when the waves reach really shallow water, of a few tens of metres or less, and particularly if funnelled into narrow bays, that the awesome power is fully developed. As each successive wave is slowed, the following ones rapidly catch up with it. With the resulting decrease in the wavelength, all the potential energy in the wave has to be dissipated somewhere. Some energy will go into the seabed, but most goes into piling the water into a huge wave, sometimes 20 metres or more high. It is this vast wall of water, weighing millions of tonnes and still moving at 10-20 km/hour when it reaches the shore, that causes the appalling devastation with which we have become only too familiar.

Predicting tsunamis is very difficult, and although there are now sophisticated early warning systems in place, particularly in the Pacific region, based on seabed instruments that record the passage of waves above them by measuring pressure changes very accurately, most victims of tsunamis have little or no warning[46]. Usually, the first indication will be the approach of visible unusually high waves. However, there are many reports of people noticing the sudden loss of water from the shoreline as if the tide had fallen rapidly. This is due to the first part of the wave train arriving being a trough rather than a crest. Under these circumstances, the inshore water is in a sense sucked back into the tsunami to form the first crest.

Although more than three quarters of all tsunamis occur in the Pacific region, understandably since this is the most seismically active part of the earth's crust, they can occur in virtually any body of water and at almost any time. They are rather rare, but certainly not unknown, in the north Atlantic where the most recent serious one occurred as a result of an offshore earthquake off the coast of Newfoundland in November 1929. The resulting tsunami was more than 7 metres high and caused the deaths of 29 people on the south coast of Newfoundland.

One hundred and seventy four years earlier, on the morning of All Saints Day 1755, the so-called Lisbon earthquake caused devastation in that city and the deaths of at least 30,000 people. In fact, like the Newfoundland earthquake, this one's epicentre was well away from Lisbon, deep beneath the Atlantic. Its effects underwater were unseen, but dramatic, resulting in a turbidite, that is a sort of submarine mud avalanche, that transported millions of tonnes of sediment across the floor of the nearby Madeira Abyssal Plain at tens of kilometres an hour, annihilating more or less everything in its path. At the same time it produced a tsunami detected in ports on the east coast of north America and in Portsmouth, England.

---

46   This was graphically demonstrated by the tsunami in Samoa and American Samoa in September 2009.

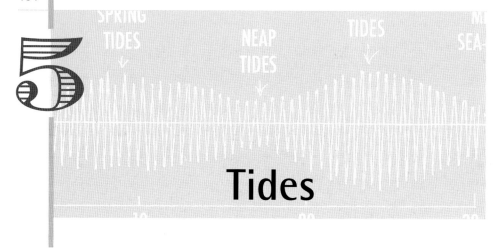

Tides

104 What are tides?

The rhythmic coastal rise and fall of the sea experienced in many parts of the world is the result of waves, in fact the longest oceanic waves on earth, which roughly follow the moon as it appears to revolve around the earth once every 24 hours (and a bit, see Q108). The pull of the moon is the main (though not the only) tide-producing force. So in a simple ocean covered world with no land masses, each point on earth would experience a high tide (that is the crest of the wave) when the moon is more or less overhead, and another one when the moon is above the opposite side of the earth. Halfway between these two times each place would experience low tide, that is corresponding to the wave's troughs. So the typical 'perfect' tide would produce two high waters and two low waters each 24 hours (and 50 minutes, see Q108 again).

Many places do, indeed, experience this 'semi diurnal' (meaning half day) type of tide. But many have a much less clear-cut pattern; for example, some places have only one high and one low water each day, called, as you might expect, a diurnal tide. Others, for example Southampton, have double high waters, with only minor ebbs between them, while Portland has the reverse, that is double low waters with only minor floods between them. Moreover, the total tidal range, that is the difference in level between the highest high tides and the lowest low tides, can range from 17 metres (more than 55 feet) (see Q109) to virtually nothing (see Q110). The variations are almost endless and are mainly the result of the presence of the earth's complicated land masses interfering with the 'simple' tidal waves that the relative motions of the moon, the earth – and the sun would otherwise produce. The next few sections attempt to explain some of these fascinating, but complex, factors.

105 What causes the tides?

The moon, stupid! I thought everybody knew that!

Well yes, we all know that the tides are caused by the pull of the moon, but it is not quite as simple as that. In fact, it is so complicated that most 'popular' explanations of the tides are such gross over simplifications that they leave far too many unanswered questions. In particular, you will often see a diagram a bit like Fig 22, showing an idealised earth and moon with the earth covered by sea with a bulge (hugely exaggerated) on the side facing the moon and a similar one on the opposite side. These bulges represent high tides and the 'thinner' bits of sea shown at right angles to the direction of the moon represent low tides. But while it is reasonably easy to imagine the pull of the moon causing the bulge towards it, it is not immediately obvious, at least not to me, why there should be a bulge in the opposite direction.

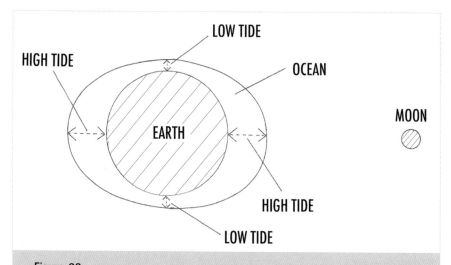

Figure 22
Simplified view of the effect of the pull of the moon on the tides. Looking down on the earth over one of the poles there are bulges in the ocean (representing high tides) both towards the moon and on the opposite side, and two thinner seas (representing low tides) at right angles to the direction of the moon.

Explaining all the intricacies of the tides adequately is far beyond this book, or me! But I'll have a crack at explaining the basics by addressing a number of fairly restricted questions, this first one dealing with the main forces involved.

There are two different types of force that produce the tides; first, gravity, the pull exerted by every celestial body by virtue of its mass (that is its weight), and second (and this is the tricky bit), centrifugal force, the tendency for a rotating body, like the earth, to throw stuff off its surface. To understand how these forces operate on the oceans we first have to remind ourselves of the relative motions of the earth and the moon in this remarkable astronomical system in which we live. And initially we will treat the earth as if it had no land but only a continuous ocean all over its surface.

The sun, as we know, is the centre of things as far as our universe is concerned. Weighing about 330,000 times as much as the earth, it exerts a huge gravitational pull. However, it is about 150 million km away from us, so its gravitational pull on earth is much less than that of the smaller, but much closer, moon which is only about 400,000 km away.

The earth and its satellite, the moon, together rotate around the sun in 365 (and a quarter) days, giving us our 'year'. But the earth and moon do some other curious things too. First, the earth spins on its polar axis once every 24 hours, giving us our 'day'. You might expect that, as it does so, the earth produces centrifugal forces, tending to throw anything on its surface out into space. And you would be right, but they have no effect on the tides. Let's see why.

The strength of any centrifugal force increases with the distance from the centre of spin. As an analogy, think how much more difficult it is to stay put on the edge of a playground roundabout than if you are near the centre. Now imagine that the earth is made up of a whole series of flat roundabouts piled on top of one another, but all spinning at the same speed. Clearly, it would be much more difficult to stand on the edge of the biggest roundabouts (that is those at the equator) than the very small ones near the poles. So the centrifugal force is at its maximum at the equator and a minimum (actually zero) at the poles, but at any particular spot on the earth's surface *it is always the same strength*. Furthermore, the centrifugal force is always in the same direction at any one place, that is along an imaginary radius of one of our piled up roundabouts. With such an unvarying strength and direction, the forces due to the earth's spinning on the polar axis have no effect on the tides.

In contrast, the centrifugal force that *does* affect the tides comes from a quite different, and much slower, motion of the earth and moon – and one that is, I fear, more difficult to grasp. Bear with me.

The earth and moon act as a single system revolving together around a common centre of mass in 27.3 days (giving us our month). Like the sun, the moon stays roughly (but not quite, see later) in the plane of the equator.

Because of the huge difference in mass between the earth and the moon, the mass of the moon being only about $1/80^{th}$ that of the earth, the shared centre of mass around which they revolve is actually inside the earth itself. If you find this difficult to grasp (and I do), imagine twirling a bunch of keys around your fist on the end of a piece of string about 20 cms long. Your fist, representing the earth, should be rotating eccentrically, and all points on its surface (and indeed, inside it) should be moving in a fairly tight circle while the bunch of keys (the moon) is following a much bigger circular path, but centred on the same point inside the earth (your fist). In our analogy, one complete revolution of the key ring round your fist represents one revolution of the moon around the earth, that is one month. But remember that during this period the real earth will have been revolving or spinning on its polar axis once every 24 hours, a motion you can't emulate with your fist. We will look at the effect of this spinning on the tides a bit later, but first let's ignore that and concentrate on the forces acting on the earth, and the sea, as a result of the eccentric movement of the earth and moon.

Since all points on the earth are, in this motion, moving in circles of exactly the same size and at the same speed, they all experience the same centrifugal force. And the direction of this force is always the same, that is exactly opposite from the direction of the moon. These centrifugal forces are shown diagrammatically as arrows all the same length (strength) on Fig 23. You can now begin to see where the sea's 'bulge' opposite the moon might come from.

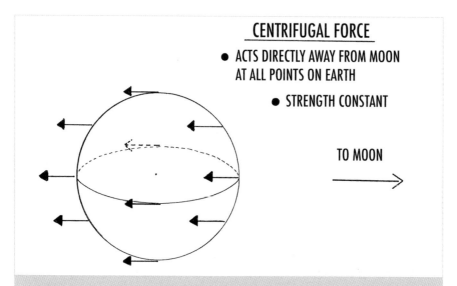

CENTRIFUGAL FORCE

- ACTS DIRECTLY AWAY FROM MOON AT ALL POINTS ON EARTH

- STRENGTH CONSTANT

TO MOON

Figure 23
The centrifugal forces acting on the earth as a result of its mutual rotation with the moon around their common centre of gravity. Note that the length of the arrows, representing the strength and direction of the force, are all the same length and all in the same direction all over the earth.

Now we have to add into the tidal equation the pull of the moon.

The gravitational pull exerted by any celestial body, like the moon, is directly related to its mass (or weight) and its distance from the object being pulled. Because the mass of the moon is only about 1/80th of that of the earth, and it is almost 400,000 km away, its pull on things on earth is pretty small. Furthermore, the pull of the moon is greater on the side facing it than on the other side, which is some 12,700 km further away. Fig 24 shows these forces pulling on the oceans as arrows just as in Fig 23, but here the arrows all point towards the moon and vary in length (that is strength), being greatest on the moon's side of the earth and least on the other side.

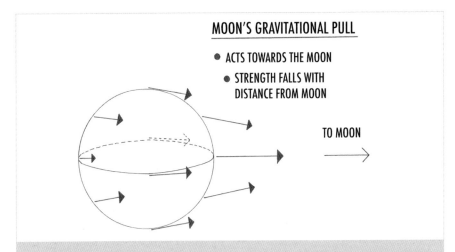

MOON'S GRAVITATIONAL PULL

- ACTS TOWARDS THE MOON
- STRENGTH FALLS WITH DISTANCE FROM MOON

TO MOON

Figure 24
Arrows representing the moon's gravity pulling on the oceans at different points on the earth's surface. They are all towards the moon, but are of different strengths; stronger on the parts nearest the moon and weaker on those parts further away from the moon.

If we add these two opposing sets of forces together we get the resultant[47] 'tide producing force', shown as another set of arrows on Fig 25. Now we can see that the balance between the centrifugal force and the moon's gravitational force produces a major pull towards the moon on one side and a force almost as big away from the moon on the other side. Halfway between the two, and at the poles, the resultant force is much smaller and directed inwards, towards the centre of the earth.

---

47 To understand what 'resultant' means, imagine two people pulling on a sledge with ropes at different angles from the sledge. The sledge will follow neither of the pullers, but will move along a line between the two ropes, the precise line depending on the relative amount of pull (that is power) being applied by the two pullers. It is the same with the tides, the net force acting on the water being the balance between the separate forces, analogous to the separate sledge pullers.

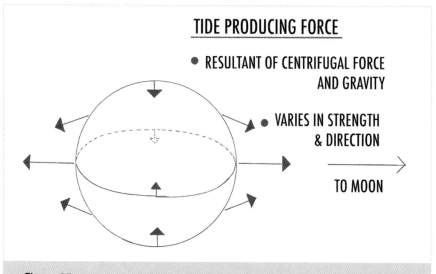

TIDE PRODUCING FORCE

- ● RESULTANT OF CENTRIFUGAL FORCE AND GRAVITY
- ● VARIES IN STRENGTH & DIRECTION

→ TO MOON

Figure 25
The net tide producing force, being the resultant of the centrifugal force and the gravitational force shown in figs 23 and 24.

But within this already quite complex situation the earth is spinning on its polar axis once a day. So now let's look again at our simple tide illustration, Fig 22, and add the earth's daily rotation (see Fig 26). We are now looking down on the North Pole so the earth appears to us to be spinning anticlockwise, that is from west to east (this, of course, is why the sun appears to rise in the east and set in the west).

Now you can see why any point on the earth's surface should experience two high waters each day (one when it is nearest to the moon and the other when it is farthest away) and two low waters when it is half way between the two. For example, take the point X on the earth's surface with the moon more or less directly overhead. As predicted by Fig 25, the tide producing force is at a maximum so the tide is high. At the same time, point Y is at a position at right angles to the direction of the moon and, again as predicted in Fig 25, the tide producing forces are at a minimum and the tide is low. Six hours later, however, both points have moved on by a quarter of a revolution, so that point X is now at position X¹, and is experiencing a low tide, while point Y is now at Y¹ (the position originally occupied by X) and is experiencing a high tide. The effect is as if the two bulges in the ocean are the crests of two very long waves travelling around the oceans from east to west (that is in the opposite direction to the earth's own direction of spin) as they are 'trying' to maintain the same position relative to the moon.

Finally, before leaving this first stage of explaining the tides we need to add

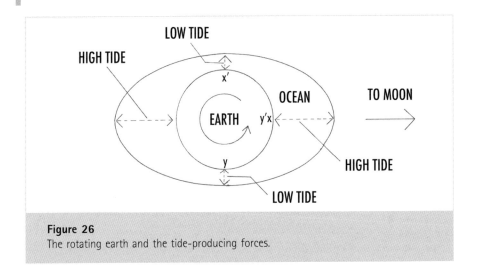

Figure 26
The rotating earth and the tide-producing forces.

one final twist. This is to point out that although the moon is by far the biggest influence on the tides, the sun is also a major player. In fact, despite its enormous distance from the earth, the sun's mass is so huge that it exerts a pull on the oceans almost half (actually about 0.46) that of the moon. And because the earth rotates on its polar axis once every 24 hours relative to the sun, the solar tides, like the lunar ones, produce two high waters and two low waters each day. The next few sections will investigate how these two sets of influences interact to produce the wide range of tidal conditions that are found on earth.

106   What are spring and neap tides?

The first thing to say is that the name spring, in this context, has absolutely nothing to do with the season. In fact, the words spring and neap in the tidal sense are a bit confusing, partly because nobody seems to know exactly where they came from or what they originally meant. *Neap* appears to be an old English word meaning goodness knows what, while *spring* is possibly derived from a Germanic word, *springen*, meaning to jump or burst. So what do they mean now?

They refer to the regular alternation, at any particular place, between tides with a big range (the spring tides) and those with a much smaller range (the neap tides). Spring and neap tides are separated by almost exactly a week, so that every month experiences two sets of spring tides and two sets of neap tides with gradual changes in the tidal range between them (see Fig 27). This provides the clue to what causes them.

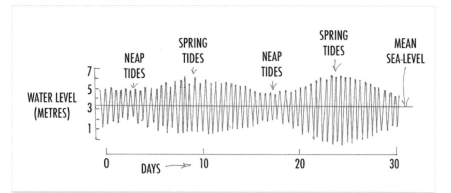

Figure 27
Typical tide record for a 30 day period, in this case from Immingham in England. The zigzag line traces the level of the water as it rose and fell over the 30 day period. This type of tide is called 'semi-diurnal' because each complete tidal cycle lasts about half a day so that there are two high waters and two low waters in each 24 hour period. However, superimposed on this basic daily pattern is another, much longer one alternating between neap tides, with a fairly restricted range, and spring tides where the range between high and low water is much greater. In this record neap tides occur around day 3 and day 17, while spring tides are around day 9 and day 23-24. See text for more details.

Q105 explained that the main tide producing forces are the pull of the moon as it changes position relative to the earth and the centrifugal force resulting from the rotation of the moon and earth together around their common centre of gravity. The result is a slight 'bulge' of the sea's surface on the side facing the moon and a similar bulge on the opposite side of the earth. As the earth rotates once every 24 hours from west to east on its polar axis, these bulges 'try' to maintain the same positions relative to the moon and therefore move around the earth as a wave from east to west[48].

However, Q105 also pointed out that the sun has an influence on the tides as well as the moon, despite its vast distance from the earth. In fact, the solar tidal force is almost half as big as the lunar one, producing a rather smaller bulge in the sea surface towards the sun and another one on the opposite side of the earth away from the sun. And as the earth spins on its polar axis these two bulges,

48   Remember that the moon is also moving round the earth from west to east, but much slower than the earth is rotating on its own axis; the earth does one revolution each day, while it takes the moon 27.3 days to complete its circuit around the earth. The net result is that *from the point of view of an observer on earth* the moon *appears* to be revolving around the earth from east to west, completing one revolution in just over 24 hours (see Q108).

like those due to the moon, produce two solar high waters and two low waters in every 24 hours. Now it doesn't take an Einstein to realise that when the lunar tide and the solar tide coincide they will produce particularly big tides, whereas when they are out of phase they will result in below average tides. Both situations occur twice each lunar month, as illustrated in Fig 28.

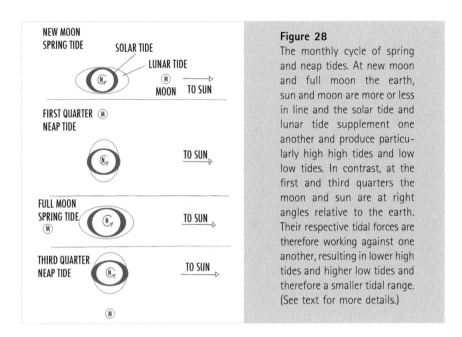

Figure 28
The monthly cycle of spring and neap tides. At new moon and full moon the earth, sun and moon are more or less in line and the solar tide and lunar tide supplement one another and produce particularly high high tides and low low tides. In contrast, at the first and third quarters the moon and sun are at right angles relative to the earth. Their respective tidal forces are therefore working against one another, resulting in lower high tides and higher low tides and therefore a smaller tidal range. (See text for more details.)

At new moon and at full moon the moon, the earth and the sun are all more or less in line and the lunar and solar tides added together produce the spring tides. When, on the other hand, the moon is in its first or third quarter, its tidal forces are acting at right angles to those of the sun. The result is smaller than average tides, the neaps. The whole cycle takes 29.5 days so we have two sets of springs and two sets of neaps every month.

107 Why are adjacent tides often of different heights?

Q106 explained how the gravitational pull of the sun interacts with that of the moon to produce the two weekly cycle of large (spring) tides and smaller (neap) tides. Figs 26 and 27, illustrating this cycle, for simplicity imagined the sun and moon staying permanently in the plane of the earth's equator and never moving north or south of this line. In such an idealised world, and with no land masses, every point on earth would experience two high waters and two low waters each day, with the biggest tidal range near the equator and decreasing towards each pole. Furthermore, there would be a regular decrease in tidal range from springs to neaps and an equally regular and smooth increase in the range from neaps back to springs. However, the world is not like that. First, it has lots of land masses that get in the way of the tidal bulges and, much more importantly for this question, the sun and moon emphatically do not stay always in the plane of the equator.

Because the earth wobbles a bit as it spins, the sun appears to move north and south each year so that it is directly above the Tropic of Cancer at about 23° North in June (to produce the northern summer) and above the Tropic of Capricorn at about 23° South in December (but see also Q81). As a result, the bulges of the solar tide also move north and south of the equator and have a small, but significant, seasonal effect on the tidal ranges.

The moon, however, moves north and south rather further – and much faster. This movement, called declination, takes the moon from about 28° North to about

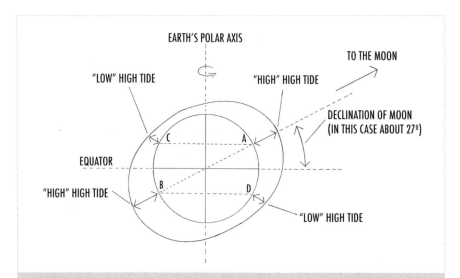

Figure 29
Variations in the tidal range caused by declination of the moon. As the earth revolves, points A and C, at similar latitudes in the northern hemisphere, each experience one tide with a high vertical range as they pass closest to the moon (position A in this fig), and a second with smaller vertical range on the opposite side of the cycle (position C in this fig). At the same time, in the southern hemisphere, positions B and D will also experience two tides of very different vertical ranges.

28° South of the equator, the whole cycle taking only 27.2 days[49]. Fig 29 illustrates one extreme of this cycle with the moon well north of the equator.

In this situation, position A, directly beneath the moon, and position B, on the opposite side of the earth, both experience large high tides. In contrast, positions C and D, which are respectively at the same latitudes as A and B but are separated by 180° of longitude, experience rather smaller high tides. Twelve hours later, as the earth has itself rotated through 180°, the positions will be reversed; localities A and B will now have rather small high tides and localities C and D will have large high tides. In exactly the same way, each of these localities will have experienced one very low low tide and one not quite so low. The net result is that all of these localities will have experienced a pair of adjacent tides with rather different ranges.

This so-called diurnal inequality in tides is particularly obvious at mid latitudes on both sides of the equator. If you look carefully at Fig 27 you will see that the two tides on individual days are often quite different, one having a considerably greater vertical range than the other. Captain Cook experienced this inequality, much to his relief, in September 1770, when his tiny ship, the *Endeavour*, ran aground on Australia's Great Barrier Reef during his first famous Pacific voyage. Despite lightening the vessel by throwing overboard some 40 tons of expendable material, the next high water failed to lift the ship off the reef, and things looked dire indeed. Twelve hours later, thanks to diurnal inequality, the next high tide rose high enough to free the vessel and the rest, as they say, is history.

108 Why do tides get later each day?

The precise timing of tides, that is the times of high and low water, can vary quite a bit at any particular locality depending upon the local conditions of, for example, wind, atmospheric pressure and topography. Nevertheless, in most places the typical tide consists of two high waters and two low waters each 24 hours – and a bit! Let's see why.

Q106 explained how the moon is the biggest influence on the tides, with the sun coming an important, but relatively poor, second. The physical forces involved 'try' to keep the two bulges in the ocean surface, causing the high

---

49   Not to be confused with the 27.3 days taken by the earth-moon system to rotate on its common centre of mass (see Q105).

tides, one on the side facing the moon and the other on the opposite side, in the same position relative to the moon. But inside this skin of water with a pair of bulges and a pair of depressions opposite one another, the earth revolves from west to east on its polar axis and makes one complete revolution *relative to the sun* in 24 hours. As it does so (and ignoring the land), every point on the earth's surface passes underneath two high waters and two low waters – almost. Almost, because while the earth is spinning merrily around, with points on the equator rushing eastwards at a staggering almost 1,700 km an hour, the moon *relative to the earth and moon's common centre of mass* is also moving in roughly the same direction, but much more slowly. In fact, the moon takes 27.3 days to make one complete revolution. In Fig 30 you can see the consequences of this.

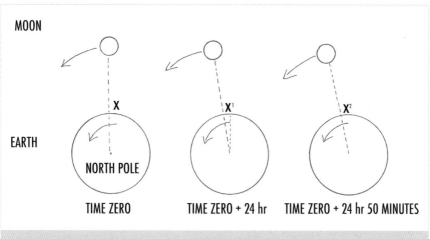

Figure 30
What a difference a day makes! The movement of the moon, and of the earth, during 24 hours . . . and a bit!

Looking down on the North Pole, Fig 30 shows point X on the earth's surface directly beneath the moon. Exactly 24 hours later X is back at the same position (marked as X[1], but by now the moon has moved on a bit. So to get back directly under the moon, position X has to wait until the earth has continued spinning a bit longer, in fact another 50 minutes because a lunar day is 24 hours and 50 minutes long. So ignoring local variations, the average time interval between two successive high waters or two successive low waters is 12 hours and 25 minutes, and the full tidal cycle moves forwards by 50 minutes each day.

109 Where are the biggest tides in the world?

As you might guess, there's a bit of a dispute, but Canada certainly claims both first and second place. The usual claimant is the Bay of Fundy, a long and narrow bay on the Atlantic coast of North America at the north-eastern end of the Gulf of Maine and bordered by the Canadian provinces of New Brunswick and Nova Scotia.

The Canadian Hydrographic Service gives a total tidal range of 17 metres (55.75 feet) at Burntcoat Head, Nova Scotia, though a high spring tide (see Q106) water level of 21.6 metres (70.8 feet) was apparently recorded in 1869 as a result of high winds and abnormally low atmospheric pressure.

The other Canadian highest tidal range claimant is Ungava Bay, an inlet off the Hudson Strait in far northern Quebec province where, says the Canadian Hydrographic Service, the extreme spring tidal range is 16.8 metres, that is, not significantly less than in the Bay of Fundy.

And the third biggest tidal ranges in the world are in the Bristol Channel where the range can be as high as 15 metres (49 feet), which is why there has been, and continues to be, interest in a Severn Barrage to harness the tidal energy to generate electricity.

110 Why are there no tides in the Mediterranean?

Well, actually it is not quite true to say that there are no tides in the Mediterranean. Although there are certainly some parts where there is no measurable tide, most Mediterranean coasts have clear tides, albeit often with ranges of no more than 20 to 30 cms. A few places, particularly near the extreme eastern and western ends and the northern Adriatic including the vicinity of Venice, can experience tidal ranges of half a metre or more, while in Tunisia's Gulf of Gabes, at the bottom of the big kink in the Mediterranean's south coast, the tidal range can be at least a metre.

Nevertheless, it is true that compared with most other parts of the world ocean the Mediterranean has remarkably small tides; so why is this? Partly it is because the Mediterranean's long and narrow east-west shape, divided as it is into two more or less isolated halves by the boot of Italy and the island of Sicily standing on a shallow ridge joining the North African coast, simply does not allow the tide producing forces, summarised in Q105, to do their stuff. But even more important is the fact that the Mediterranean is almost entirely cut off from the rest of the oceans except for the very narrow and shallow Strait of Gibraltar. All tides involve extensive, and sometimes massive, movements of water, resulting in tidal currents that often flow at several knots; but there are limits. For the Mediterranean to have tidal ranges measured in metres rather than centimetres, the transport of water from and to the open Atlantic during each tidal cycle would have to be enormous – and the Strait of Gibraltar simply isn't big enough; hence, hardly any tides[50].

111 What are tidal bores?

A tidal bore is a curious phenomenon in which the leading edge of the incoming tide produces a wave, or series of waves, that travel up a river or narrow inlet against the outflowing current. They are true tidal waves and are not to be confused with tsunamis (Q103) which are often called tidal waves, but have nothing to do with the tides.

Bores are rather rare, occurring in only a few locations worldwide, almost always where the tidal range is large and where the tide is funnelled into a shallow, narrowing river from a relatively wide bay.

The largest bore in the world is on the Qiantang River in China, where a wave up to 9 metres (30 feet) high travels up-river at speeds approaching 40 km (25 miles) an hour.

The mouth of the Amazon also has a bore, though in this case no more than 4 metres (13 feet) high and travelling at 15-25 km (9-16 miles) per hour.

North America has rather few bores but, not surprisingly, several of those it has are in the Bay of Fundy region, associated with the very large tides (see Q109). The highest bore in North America, more than 2 metres high, was in the Petitcodiac

---

50 William Shakespeare conveniently ignored this fact when he wrote *Julius Caesar*. Before the Battle of Philippi he has Brutus uttering the following famous words:

'There is a tide in the affairs of men
Which, taken at the flood, leads on to fortune;
Omitted, all the voyage of their life
Is bound in shallows and in miseries.'

As a good Roman born and bred, it would surely never have occurred to Brutus that the 30 cm high tides with which he was familiar could have any such effect. Thank goodness Will didn't let this get in the way of the birth of this cracking tidal quote!

River, emptying into the Bay of Fundy, but it has been more or less eliminated in recent years by siltation and the construction of a causeway.

Europe has quite a lot of bores, and the name itself comes from an Old Norse word meaning wave or swell, though the French usually use the term *mascaret*.

Britain's most famous bores are on the River Trent[51] and the Severn, where the bore can be up to 2 metres high and frequently attracts the attention of surfers and canoeists.

---

51   Where it is known as an eagre or aegir, in turn possibly from Aegir, a Scandinavian god of the sea.

The Oceans And Global Warming

112 Are the oceans getting warmer?

Yes, they are, and don't let any global warming sceptic convince you otherwise. Mind you, when these sceptics claim that ocean warming is nothing new they are, of course, quite right. The oceans have been warming ever since the end of the last ice age. However, the rate of warming in the last few decades is unprecedented.

Measuring changes in average temperatures of the air or the oceans is not easy. Reconstructing past temperatures is even more difficult, despite the availability of some very clever and sensitive techniques for doing so.

A classic recent example is the controversy over the famous so-called 'hockey stick curve', the name given to the reconstruction of Northern Hemisphere mean temperatures over the past 1,000 years that first appeared in 1998 (see Fig 31). The hockey stick epithet referred to its shape, with a relatively straight and level section from 1000 AD to about 1900 and then a rapid upward curve at the end. The curve received a great deal of criticism, mostly from other scientists arguing about the statistical methods employed in its production. In the intervening years the data have been re-examined and, while there are still serious concerns about some aspects, the overall result suggesting that global temperatures have risen faster in the last century than at any time in the previous 1,000 years or more appears to be vindicated (see an excellent Wikipedia entry under 'Hockey stick controversy'). Unfortunately, the curve was used rather uncritically by a number of enthusiastic global warming prophets (including Al Gore in his film *An Inconvenient Truth*), providing sceptics with ammunition to challenge the whole climate change argument.

Estimates of past changes in seawater temperatures are subject to the same uncertainties and therefore criticisms, but this is not true of measurements made over the last 50 to 100 years. Scientists now have instruments that can measure ocean temperatures to hundredths of a degree many times each second as they sink through the

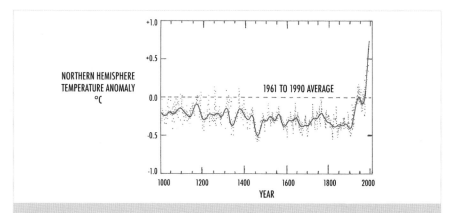

Figure 31
The so-called Hockey-stick curve, redrawn from the IPCC 3rd Assessment Report, 2001. The curve shows the estimated changes in the mean temperatures in the Northern Hemisphere over the last 1,000 years relative to the mean temperature between 1961 and 1990 indicated by the dashed line across the middle of the graph labelled 0°C. The wavy solid line traces departures from this 1961-1990 average smoothed by averaging over 40 years, those parts below the dashed line indicating cooler periods and those above the line indicating warmer ones. The cloud of dots around the solid line indicates year to year variations.

water column. In recent decades they have also had the technology, using satellites, to measure sea surface temperatures over the whole of the oceans in the space of a few hours or days, and to repeat the exercise time after time. So when the International Panel on Climate Change (IPCC) say, as they do in their latest (2007) assessment, that 'over the period 1961 to 2003, global ocean temperature has risen by 0.1°C from the surface to a depth of 700 metres', they do so with a great deal of confidence. If you think a rise of only one tenth of a degree is small beer and nothing to get excited about, remember that it takes more than 1,000 times as much energy to raise the temperature of a cubic metre of water through one degree as it does for the same volume of air.

Furthermore, if you consider that this heating has resulted from the oceans absorbing 90% of the excess heat in the atmosphere due to higher greenhouse gas levels, compared with only 3% heating up the air and 7% heating land and ice, you will realise what a vast amount of heat it represents. In fact, says the IPCC, it is equivalent to absorbing energy at the rate of one fifth of a watt for every square metre of the entire earth's surface! Remember also that the earth's climate and weather are controlled largely by exchanges of heat between the planet's two great fluid systems, the oceans and the atmosphere. In the short term the oceans may absorb all this energy without causing too much trouble, give or take a few more intense tropical storms and rising sea levels. But sooner or later they will hand it back to the atmosphere, because, as sure as eggs are eggs, a warmer ocean means a warmer atmosphere.

Finally, the average rise in temperature of 0.1°C of the surface oceans hides the fact that some parts of the ocean have actually cooled and some parts have become warmer by much more than one tenth of a degree. They also have to be seen against a background of week to week and year to year temperature variations associated with phenomena such as El Niño. A similar, but rather less dramatic oscillation (called the Pacific Decadal Oscillation or PDO) occurs in the north Pacific and an even longer term one, the Atlantic Multidecadal Oscillation (AMO), in the North Atlantic. It appears that the net result of all these separate oscillations is that over the next few years the earth may see a slight *cooling* rather than warming, masking the more general long-term trend. Unfortunately, as the then Prime Minister, Harold Wilson, said many years ago, 'a week is a long time in politics'. Politicians generally have rather short horizons, rarely extending much beyond the next election. Consequently, many scientists fear that climate change sceptics will seize on these short term 'blips' and use them in attempts to dissuade politicians from taking unpopular and potentially expensive measures to counter global warming. For all our sakes, let's hope that wiser counsels prevail.

113 Is the sea level rising?

Yes it is, and it has been for the last 15-20,000 years! Since the peak of the last ice age, about 18,000 years ago, sea level has risen about 130 metres, so that before this time most of the world's continental shelves were above water.

Sea level can *appear* to rise or fall relative to the land as a result of geological activity causing vertical movements of the land itself. But true sea level changes, due to changes in the volume of the oceans, result from two main causes: first, changes in the amount of water locked up in the earth's ice caps and glaciers; second, expansion and contraction of the seawater itself as it gets warmer or colder.

Both of these factors have contributed to the sea level rise since the last ice age, the average rate over this period being about 7mm/year. But most of the rise took place more than 6,000 years ago. For the last 3,000 years or so, sea levels have risen very slowly, at no more than 10 to 20 cm every 1,000 years (that is 0.1-0.2mm/year), until about the beginning of the 19th century. Since then, things have changed dramatically. Estimates from different sources such as tide gauge measurements and, more recently, satellite altimetry, vary a bit, but all agree that the annual rate during the 19th and 20th centuries has been at least ten times higher than previously. Best estimates for the last 130 years or so suggest an average rate of a little less than 2mm/year with a total rise through the 20th century of almost 20 cm. There is some indication that the rate has accelerated recently but this is uncertain. For example, during the period 1993 to 2003 the rate of sea level rise appears to have been a little over 3 mm each year, rather higher than the average rate. But sea level changes are very variable both in time and space, so that in some places the level may actually fall, at least temporarily. Consequently, it is difficult to say whether the recent apparent acceleration is simply a blip or will be continued. Even disregarding any such acceleration, however, estimates of the expected sea level rise during the 21st century range from as little as 9 cm and as much as 88 cm, with a mean value of 48 cm. At the lower end of this range the effects would be significant; at the upper end they would potentially be very serious. Many coastal areas would be more susceptible to coastal erosion, storm surge flooding and deterioration in the quality of surface ground water and therefore agricultural production. In some areas, such as low lying islands in the Pacific and Indian Ocean, the effects would be even more direct and dramatic. In areas of high human population density the results could be catastrophic. In the Bay of Bengal, for example, it is estimated that a sea level rise of 40 cm would result in 7-10 million refugees.

There are still lots of uncertainties about sea level rise, including the contribution made by melting ice caps on the one hand, or thermal expansion of the water on the other. For the period between 1961 and 2003, the International Panel on Climate Change (IPCC) most recent report estimates that thermal expansion contributed only 0.4 mm of the average of 1.8 mm annual rise during this period. Puzzlingly, the estimated contribution from ice melts during this period isn't enough to make up the difference. In contrast, in the period from 1993 to 2003, thermal expansion explains about 1.6 mm rise each year, while loss of mass from glaciers, ice caps and the Greenland and Antarctic ice sheets accounts for a further 1.2 mm annual rise. Together, these are not far short of the observed 3.1 mm average annual rise based on satellite altimeter readings, providing confidence boosting, but somewhat worrying, reassurance of agreement between the sums.

However, serious though these figures are, they are insignificant compared with what could happen if the Arctic and Antarctic ice caps were to melt completely. Melting of the entire Greenland icecap would result in a sea level rise of about 7 metres while melting the Antarctic cap would raise the sea level by more than 60 metres. Fortunately, even in the worst global warming scenarios such changes would take several millennia.

114 Do the oceans absorb carbon dioxide?

Yes, they do. In fact, they have absorbed almost half of the anthropogenic (that is, man-produced) carbon dioxide since the beginning of the industrial revolution. If they hadn't, the level of carbon dioxide in the atmosphere, and therefore the greenhouse effect, would be considerably greater than it is today. But whether the oceans can continue to do this CO_2 soaking up job for much longer is very debatable.

Carbon dioxide is never very abundant in the atmosphere. However, in terms of its greenhouse effect it punches much heavier than its weight, so to speak, so that variations, even in tiny amounts, are crucial. During ice ages, CO_2 levels are generally pretty low, around 180-190 ppm (parts per million) (that's 0.018 to 0.019%). During warm interglacial periods, like the one we are in now, the CO_2 level typically rises to around 290 ppm, and from about 20,000 years ago until the end of the 19th century the levels ranged between about 260 and 290 ppm. But since then it has rocketed to 383 ppm in 2007, and it is still rising at between 2 and 3 ppm every year. The IPCC's most recent report says that it is 'more likely than not' that between 1750 and 1994 something like 42% of the emitted CO_2 was taken up by the oceans, whereas between 1980 and 2005 this proportion dropped to about 37%. If this result is confirmed, it would suggest, rather worryingly, that the oceans' capacity to absorb CO_2 is wearing a bit thin.

So far the total inorganic carbon content of the oceans has increased by more than 100 gigatonnes, that is one hundred thousand million tonnes. This is a huge figure, but in view of the huge size of the oceans it is probably not, alone, particularly significant.

On the other hand, it has some very important consequences, particularly through its effect on the acidity of seawater. The chemistry of carbonate ions in water is extremely complex, but put over-simply, CO_2 dissolved in water produces carbonic acid. This is a pretty weak acid but one that has reduced the average pH of the surface waters of the ocean by 0.1 pH units since the industrial revolution (from about pH 8.179 to pH 8.104). This may not seem much of a change, but it has been enough to make life difficult for lots of creatures that use calcium carbonate (basically chalk) in their skeletons, such as molluscs, corals and coccolithophorid algae. The drop in seawater pH has made things just a little bit tougher for these chalky organisms to make a living, with the result that the area of the earth's warm oceans that are 'coral friendly' has decreased significantly in recent years, causing lots of reefs to die off. Sad though this is, in the overall scheme of things the effect on coccolithophorids may be even more serious. Coccolithophorids make up one of the most important constituents of the phytoplankton and play a crucial role, not only in the overall productivity of the oceans, but in the drawdown of carbon dioxide from the atmosphere, thus helping to alleviate the greenhouse effect. Increased ocean surface acidity may reduce coccolithophorid abundance and, indirectly, accelerate global warming.

Man and the sea

115 How deep have men been in the oceans?

The deepest dive so far recorded took place on 23 January 1960 in the Challenger Deep in the southern part of the Marianas Trench in the south-western Pacific, when Jacques Piccard and Don Walsh (Ltnt, US navy) reached a depth of about 10,900 metres or 35,800 feet in the bathyscaphe *Trieste*, at that time owned by the US Navy.

The *Trieste* had been designed by Jacques Piccard's father, the Swiss scientist Auguste Piccard, and had been launched in the Mediterranean in 1953. It had a main hull, or 'gondola', some 15 metres long and consisting principally of buoyancy tanks containing 85m³ of gasoline (petrol), together with water-filled ballast tanks at either end and 9 tons of iron pellet shot to carry the vessel to the bottom. The shot was held in place by electromagnets so that, in the event of a power failure, it would be released and the *Trieste* would return to the surface.

The two passengers were housed in a 2.16 metre (6.5 feet) diameter pressure sphere with a single cone-shaped plexiglass (that is perspex) window to the outside.

The descent took 4 hours and 48 minutes and, after spending about 20 minutes on the bottom, *Trieste* returned to the surface in 3 hours 15 minutes.

116 Why do scuba divers have to breathe out when they return to the surface?

The demand valve connecting the scuba diver's high pressure air tanks with his breathing tubes always provides him with air at the same pressure as the surrounding water, whatever the depth. Consequently, as he breathes in and out, his lungs fill up and empty just as they would at the surface – so long as he doesn't change his depth very suddenly. However, if he dives deeper and holds his breath he will soon feel pressure on his chest as the volume of air in his lungs decreases and his rib cage collapses. If, on the other hand, he holds his breath and moves upwards he will feel pressure from inside his chest as his lungs get bigger and bigger. Eventually, if he still ascends without breathing out, his lungs may actually burst like an over-inflated balloon, releasing air bubbles into the chest cavity. This is a potentially very serious condition called spontaneous pneumothorax (not to be confused with the bends, a quite different condition dealt with in Q66).

Interestingly, this problem of expanding air becomes more critical the nearer the diver is to the surface (see also Q30). Imagine, for example, what happens if a scuba diver started his ascent at the huge depth of 90 metres where the pressure would be 10 atmospheres. If he holds his breath and moves up 10 metres to 80 metres depth, the pressure would decrease to nine atmospheres and his lung volume would have increased by one tenth. Therefore, in order to keep his lungs at the same volume during this upward movement the diver would need to breathe out only about one tenth of the contained air. In contrast, if he started his ascent at the modest depth of only 10 metres, where the pressure is two atmospheres, as he came to the surface moving exactly the same vertical distance (10 metres), the pressure would drop to one atmosphere and his lungs would double in volume. So this time he would have to get rid of half of the air in his lungs to keep the volume the same.

117 Does drinking seawater drive you mad?

Probably not, but drink too much and it will certainly kill you.

Human beings all need water and salt to survive; both are in seawater, so why is seawater dangerous? The problem is a matter of balance. Whereas we need quite a lot of water (in total about 2.5 litres/day from all sources), we need only a small amount of salt (about 500 mg/day). As we are constantly being told, most of us in the developed world take in far too much salt for our own good, and consequently we need to drink much more water in order to enable our kidneys to flush it out of our systems. And here is the rub. The saltiest urine that human kidneys can produce is slightly *less* salt than seawater.

Seawater contains about 35 grams in every litre, so if we drank our daily 2.5 litres as seawater we would be taking in about 170 times as much salt as we need[52], and far from quenching our thirst, it will actually make us even more thirsty. This is because drinking it causes acute dehydration, that is loss of water from the blood and body tissues.

Human blood is quite salty, but only about 0.9% by weight[53] compared with the 3.5% or thereabouts for seawater. If you drink seawater it causes an increase in the amount of sodium in the blood and this, in turn, promotes sodium excretion by the kidneys. But since the kidneys can't produce very salty urine, if you carry on drinking seawater the concentration of sodium and other salts in the blood will continue going up. This will essentially 'suck' water out of the body's cells and tissues, causing acute dehydration and interfering with all sorts of bodily functions including nerve conduction, possibly leading to seizures and heart failure. So you may not be mad, but you certainly won't be too pleased!

Of course, drinking a small amount of seawater, for example when you are swimming in the sea, will cause no significant harm, particularly if you take in reasonable quantities of fresh water soon afterwards. And as you might expect, you can drink much larger quantities of less salty or brackish water without doing much harm. Indeed, some particular groups of people are capable of drinking large quantities of salty water that would be intolerable to most of us. I remember many years ago, during a working visit to Bahrain in the Arabian Gulf, being told that the local pearl fishermen who stayed at sea for two or three weeks at a time, took no freshwater supplies with them. Instead, once they reached their chosen work site, the first dives would be to fill goatskin vessels with water seeping through the sea floor from underground freshwater aquifers. This water inevitably became mixed with at least a little seawater before it could be collected in the goat skins, and was consequently quite salty by the time the men drank it. But because they were habituated to it, no significant harm appears to have been caused.

In contrast, most people are very sensitive to drinking salt water and modern

---

52 Quite apart from the fact that, in addition to 'ordinary' salt, or sodium chloride, sea water salt contains lots of other ingredients that we don't need, and can do harm.

53 Consequently, this is the strength of 'normal saline', the fluid that is dripped into the bloodstreams of human patients that have become dehydrated from whatever cause.

survival manuals for seafarers generally advise against drinking any seawater at all. This is apparently backed up by the results of a number of situations where wreck survivors have spent long periods on life rafts. Despite the accumulation of toxins in the absence of sufficient water to allow the kidneys to flush these from the body, it seems that survival rates are significantly higher for those unfortunate shipwrecked mariners who resist drinking any seawater than for those who succumb.

In this context, older readers may remember the name Alain Bombard from the 1950s[54]. Bombard (who died at the age of 80 in 2005) was a French physician who became famous in 1952 for sailing solo across the Atlantic in a small boat with virtually no stores of food or water. Bombard's objective was to try to demonstrate that shipwrecked seafarers could survive long periods at sea with the minimum of help. As far as water intake was concerned, his thesis was that drinking small amounts of sea water each day would enhance survival rates and cause no significant lasting damage. But, in addition, he tried to demonstrate that fish tissues and abdominal fluids contained sufficient more or less fresh water to supplement the salt water intake.

When Bombard arrived at Barbados in December 1952 after more than four months at sea he was in pretty poor condition; but he had survived, and his exploits earned him considerable acclaim and the gratitude of many seafarers. Nevertheless, he was not without his critics and sufficient doubt has been expressed about the truth of some of his claims that he remained something of a controversial figure throughout the remainder of his life.

Finally, let's finish with the most famous quote ever about the potability of seawater, from Samuel Taylor Coleridge's *Rhyme of the Ancient Mariner*. Having killed the albatross the bad luck begins with the ship becalmed in a tropical hell:

Day after day, day after day,
We stuck, nor breath nor motion;
As idle as a painted ship
Upon a painted ocean.

And then the killer verse; famine in the apparent midst of plenty:

Water, water, every where,
And all the boards did shrink;
Water, water, every where,
Nor any drop to drink.

---

54   See *The Bombard Story*, by Alain Bombard, published in 1953 by André Deutsch Ltd.

118 Who owns the oceans?

If you had asked this question 70 years ago the answer would have been pretty simple: apart from very narrow strips around coastlines, no one does! But since the end of the Second World War things have become a good deal more complicated.

In the 1930s the maritime legal situation was much the same as it had been for the previous 300 years. There were few established agreements, but the generally accepted situation was that each maritime nation, that is one with a coastline, could claim jurisdiction over a strip of the sea three miles wide extending all round its coast. The origin of the three mile figure is unclear, though it is often claimed, possibly correctly, to have been based on the range of early 17th century canon. If so, the thinking was that with such guns mounted along the shoreline a nation could control all activities in these 'Territorial waters'. Beyond these limits the 'high seas' belonged to no one. Although the implication was that anyone could use these high seas, in practice they were always controlled by the most powerful maritime nations or, in some areas, powerful non-national groups like pirates[55].

However, as technology improved through the eighteenth and nineteenth centuries, the range of guns greatly exceeded three miles, making the original argument for a three mile limit rather irrelevant. It was nevertheless maintained, mainly because the narrower this band was, the more of the ocean could be controlled by the major maritime nations, initially Britain, France and Russia and later the USA and Germany.

In the early years of the 20th century, pressure for change built up within the less powerful nations, but nothing constructive was achieved. However, things changed rapidly after the end of the Second World War. First, in 1945 the USA, anxious to secure its oil supplies, laid claim to the mineral resources beneath its Continental Shelves, and it also claimed exclusive rights to the fishery resources in the overlying waters.

Over the next few years a number of nations followed the US example and extended their territorial water claims to 12 miles, while four South American nations, Argentina, Chile, Peru and Ecuador, claimed unprecedented 200 nautical mile wide exclusive fishing grounds. But events in the North Atlantic were even more dramatic. Iceland increased its territorial water claim to 4 miles in 1950, 12 miles in 1958, 50 miles in 1972 and finally to 200 miles in 1975, all in response to what it saw as decline in its fishing stocks as a result of overexploitation by foreign

---

55   Just how easy it is for renegade groups to wreak havoc on the high seas has been graphically illustrated by the recent upsurge in piracy in the Indian Ocean.

fishing vessels, particularly British. These led to the three so-called cod wars. Clearly things were getting out of hand and the United Nations became involved.

In 1956 the United Nations Conference on the Law of the Sea, known as UNCLOS for short, was established – and remains in existence to this day. This is not surprising, because international maritime law, which is UNCLOS's remit, is hugely complicated and there will always be matters for lawyers to argue about.

In fact, the problems were so complicated that UNCLOS's first two attempts (in 1958 and 1960) to get some sort of general agreement on who had the right to do whatever in the oceans failed totally. But UNCLOS persevered and what became known as UNCLOS 3 sat from 1973 to 1982, finally producing a convention that is in broad use today (see Fig 32).

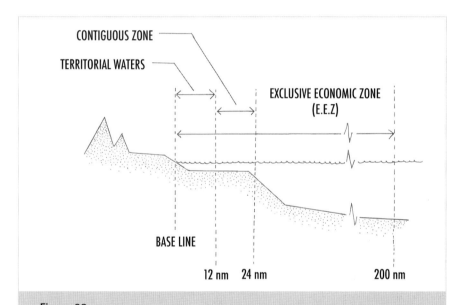

Figure 32
Cross section of a hypothetical piece of land and the adjacent ocean to illustrate the conclusions of UNCLOS 3. The limit of the Contiguous Zone, at 24 nautical miles from the baseline, is shown here as just beyond the edge of the Continental Shelf. However, the boundaries of the various zones are not directly related to the features of the underlying sea floor, so that in any particular case the shelf edge might be inside the Contiguous Zone limit, as shown here, or well outside it.

So what does it say? Well, it defines a series of zones parallel to a nation's coastline over which it has various degrees of authority. But first, the convention defines how each nation shall establish the *baseline* from which these zones are to be measured. This is extremely complicated and detailed because it has to deal with different sorts of coastal indentations like bays and river mouths. And it also makes recommendations on the thorny issue of how a coastal state should deal with offshore islands and whether or not they can be used to extend the baseline.

Having established the baseline, UNCLOS 3 states that in a band extending 12 nautical miles seaward of it (called *Territorial Sea*) the coastal state has more or less total jurisdiction and control over any resources, with the proviso that the vessels of other nations are given the right of 'innocent passage' through this zone. This also provides for the passage of naval vessels through strategic straits, like the Straits of Dover, so long as this passage is not 'prejudicial to the peace, good order or the security' of the coastal state.

Second, the Convention establishes a second 12 nautical mile wide zone outside the Territorial Sea, the *Contiguous Zone*, within which it could extend its jurisdiction over activities like smuggling or illegal immigration.

Finally, the Convention defined an *Exclusive Economic Zone* or EEZ extending 200 nautical miles from the baseline (that is including the Territorial Sea and Contiguous Zone), within which the coastal state has sole exploitation rights over all (that is living and non-living) natural resources.

This all seemed, and indeed was, relatively straightforward. But UNCLOS 3 added a couple of little twists. The first was to say that, notwithstanding the above conditions, a coastal state could claim jurisdiction over all the mineral and other non-living resources in the subsoil of its Continental Shelf (see Q13), since this is geologically speaking a natural extension of the nation's land territory. But some continental shelves are much wider than 200 nautical miles, so the somewhat infamous Article 76 of the convention said that, under such circumstances, the coastal state could extend its EEZ limit to beyond the 200 nautical mile line, but never beyond 350 nautical miles. Oh, or 100 nautical miles beyond the 2,500 metre isobath (that's a line joining points at a depth of 2,500 metres on the continental slope).

The implication of all this was that beyond these designated zones the High Seas, and their resources, were not under the control of any individual nation. But now came the second little twist, what became known as Part XI. This dealt with the mineral resources on the seabed outside any state's EEZ (for example oil, gas or manganese nodules) and established an independent International Seabed Authority (ISA) under the auspices of the International Maritime Organisation (IMO), with the remit of authorizing seabed exploration and mining in these areas and collecting and distributing the resulting royalties.

As with all such international conventions, UNCLOS conventions come into force only after a predetermined number of participating nations have ratified them, and even then they are binding only for those who have ratified. So if you don't ratify, you don't have to abide by the 'rules' even if you participated in formulating them! That's international politics for you.

In the case of UNCLOS 3 there were 155 participants and 60 ratifications were required before it came into force. This happened on 16 November 1994, one year after ratification by Guyana. Not surprisingly, Iceland had ratified early (the 21st to do so, in 1985), but in 1994 many major maritime nations had still not ratified it, though many have subsequently. For example, Australia (number 65) ratified in 1994; France (87), Ireland (97) and Norway (98) in 1996; Russia (114) and the UK (118) in 1997, and Denmark (146) in 2004. Among the 24 signatory countries that have still not ratified the convention are Afghanistan, Ethiopia, Iran, Korea, Libya, the United Arab Emirates – and the United States!

The US is, of course, by far the most important of these non ratifying countries, and the reason is basically Part XI (see above). The USA has no problem with the remaining parts of the convention and, indeed, considers them binding. But it considered Part XI as unfavourable to America's economy and security, and more favourable to the economic systems of the communist states.

With the break-up of the communist block in the late 1980s, and the decline in the perceived demand for seabed minerals in the 1990s, the problems posed by Part XI decreased somewhat in recent years. Nevertheless, the principle is still a matter for discussion within US politics, with a majority of the US senate in favour

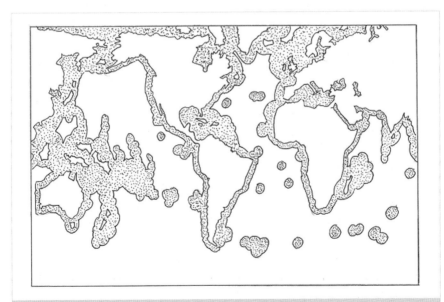

Figure 33
The world's Exclusive Economic Zones under UNCLOS 3. Note that because this is on a Mercator projection (see Q84), the high latitude areas are greatly exaggerated. Nevertheless, note also that a relatively enormous amount of the deep sea comes under the jurisdiction of tiny independent island nations in the central Pacific.

of ratification, but with a significant faction of conservative Republicans adamantly opposed. Watch this space and see what the new administration does!

Now just look at Fig 33 to see how much of the world's oceans come under the jurisdiction of one nation or another. This is a Mercator projection so that, as is always true of this projection (see Q84), areas at high latitudes are greatly exaggerated. Nevertheless, it is clear that a vast amount of the oceans are now technically 'owned': in fact, rather more than 40% lies within an Exclusive Economic Zone and, as a corollary, less than 60% is now classified as High Seas. Furthermore, a huge proportion of these EEZs surround tiny islands in the central Pacific, some being outlying parts of big nations like France, the UK and the USA. But quite a lot of them belong to recently independent island states like Fiji, Kiribati, Tuvalu and Western Samoa with tiny populations. So in theory, at least, each citizen of these little states 'owns' much more deep sea than citizens of, for example, the UK, the USA or Russia, and infinitely more than the citizens of the many countries with no coastlines at all. When and if the deep sea potential resources like manganese nodules come into their own, these differences could start to become politically very significant. Watch this space, too!

INDEX

Please note, numbers refer to questions, not pages

Abyssal hills: 17
Abyssal plains: 13; 14; 15; 16; 17; 103
Acidity: 114
Agonic lines: 82
Aleutian Ballad (fishing vessel in the TV programme *Deadliest Catch*): 102
Amazon, River: 111
Amundsen, Roald (polar explorer): 81
Anadromous (fish): 53
Ancient Mariner, Rhyme of the: 117
Antarctica: 3; 9; 10; 11; 20; 21; 22; 28; 42; 47; 56; 59; 61; 81; 82; 84; 113
Arctic: 9; 10; 11; 21; 28; 33; 56; 81; 82; 113
Asthenosphere: 22
Atmosphere: 1; 2; 5; 7; 30; 114; 116
Autotrophs: 27

Baleen (see whalebone)
Basking shark: 50
Bay of Fundy, Canada (record tide claimant): 109; 111
Beaufort, Sir Francis (Britsh Hydrographer): 94
Bioluminescence: 27; 37
Blue-green algae (see Cyanobacteria)
Breakers: 100
Bristol Channel (large tidal range): 109
Bulbous bows: 87; 89

Catadromous (fish): 53
Carbon dioxide: 28; 23; 25; 27; 28; 66; 114
Cephalopods: 40;
Challenger, HMS (Royal Naval survey vessel): 6; 31; 115
Challenger Deep: 6; 115
Chemosynthesis: 18;

Chlorophyll: 8; 25; 28
Circulation: 4; 21; 97; 98
Climate change: 112; 113
Cnidaria: 33; 34
Coccolithophores: 8; 27; 114
Coleridge, Samuel Taylor: 117
Compensation depth: 24; 28
Contiguous zone: 118
Continental Shelves: 13; 17; 22; 44; 118
Continental Slopes: 14: 15
Cook, Capt James (Royal Naval explorer): 94: 107
Corals: 114
Coriolis Force: 19; 20; 28; 88
Crustaceans: 31; 33; 34; 36; 38; 39; 40; 42; 43; 44; 45; 46; 47; 48; 63
Currents: 11; 17; 19; 20; 23; 26; 28; 31; 32; 34; 85; 93; 98; 102; 110; 118
Cyanobacteria: 27; 39

Dalrymple, Alexander (first Royal Naval Hydrographer): 94
Deadliest Catch (television programme)[56]: 102
Diatoms: 27; 32; 39
Dinoflagellates: 27; 39
Displacement: 88
Discovery RRS (British research vessel): 101
Dolphins: 50; 52; 55; 58; 62; 63; 64; 65; 66; 68; 71; 72; 76; 77
Dublin Bay prawn: 44; 45

Earthquakes: 103;
Echinoderms: 31; 48
El Niño: 112;
Endeavour (HMS), Capt. Cook's ship: 107
Exclusive Economic Zone (EEZ): 118

Ferrel cells: 98
Fishing: 118
Fitzroy, Robert (founder of the Met Office): 96
Flying fish: 28; 51; 52
Food poisoning: 39
Fucoxanthin: 25

Geology: 5; 31
Gimbals: 83
Global warming: 21; 28; 112; 114
Glomerulus: 53
Gondwanaland: 3
Gore, Al (Albert Arnold) (American politician and environmental activist): 112
Gulls: 54; 55; 56;
Gyrocompass: 82; 83

Hadley Cells: 98
Hockey-stick curve (controversial graph of global temperatures): 112
High seas: 118
Hydrothermal vents: 17; 18; 28; 31

Ice: 9; 10; 11; 12; 59; 61; 113
Icebergs: 9; 10; 11; 12;
IMO (International Maritime Organisation): 87; 118
Insects: 35; 36
IPCC (International Panel on Climate Change): 112; 113
ISA (International Seabed Authority): 118

Jellyfish: 26; 33; 34; 35; 36

Knot: 21; 51; 85; 93; 101; 110

Lading: 86
Larboard: 86
Latitude: 8; 19; 20; 22; 24; 27; 28; 32; 80; 81; 82; 84; 97; 98; 107; 118
Light: 8; 10; 18; 23; 24; 25; 27; 28; 29; 32; 37
Lion's mane jelly: 33
Lisbon earthquake: 103
Lithosphere: 22
Longitude: 19; 81; 84; 107

Magnetic North: 82; 83

Magnetic variation: 82
Manganese nodules: 5; 18; 118
Mariana Trench: 6; 7; 31
Mercator projection: 84; 118
Mid-ocean ridges: 17
Molluscs: 31; 32; 39; 40; 41; 42; 114
Moon jelly: 33

Nautical mile: 17; 19; 28; 80; 81; 82; 85; 91; 92; 94; 95; 102; 118
Neap tides: 106
Norway lobster: 44; 45;

Ocean skaters (insects): 36
Oceanography 31;
Octopus: 40

Pangea: 3
Panthalassa: 3
Penguins: 47; 55; 58; 59; 60; 61
Petitcodiac River (tidal bore): 111
pH (acidity): 114
Phytoplankton: 8; 13; 24; 25; 26; 27; 28; 31; 32; 39; 43; 47; 114
Photosynthesis: 18; 23; 24; 27; 28;
Piccard, Jacques and Auguste: 31; 115
Picoplankton: 27
Plankton: 8; 13; 24; 25; 26; 27; 28; 29; 31; 32; 33; 39; 43; 47; 52; 114
Plate tectonics: 3; 22
Plimsoll Line: 87
Polar Cells: 98
Pond skaters: 36
Porpoise: 64; 72
Port and starboard: 86
Portuguese man-o-war: 34
Posh: 86
Prawns: 43; 44; 46
Precession: 83
Pressure: 7; 9; 18; 28; 30; 31; 50; 52; 65; 66; 89; 93; 95; 98; 101; 102; 103; 108; 109; 115; 116

Queen Elizabeth 2 (Cunard vessel): 102
Queen Mary (Cunard vessel): 102
Queen Mary 2 (Cunard vessel): 88
Quiantang River, China: 111

Ramapo, USS: (record of largest wave): 101

Rockall: 96; 101
Rogue waves: 102
Rorquals: 63; 68; 74; 75
Ross, James Clark (Royal Naval
    explorer): 82

Salinity/saltiness: 4; 5; 7; 22; 32; 48;
    117
Scampi: 44; 45
Scyphistoma: 33
Sea areas: 95; 96
Sea
    -Mediterranean: 3; 31; 44; 45; 56;
        110; 115;
    -North: 4; 13; 27; 28; 95; 96
Seabirds: 28; 36; 55; 56
Seagulls: 55;56
Seafloor spreading: 3; 22
Seals: 26; 32; 47; 55; 58; 63; 65; 71
Seamounts: 17
Sea striders (insects): 36
Seaweeds: 23; 24; 25; 31; 33;
Sediment: 5; 17; 18; 22; 27; 28; 31;
    32; 40; 43; 100; 103
Severn, River (tidal bore): 111
Sharks: 42; 47; 50; 52; 53
Shelf break: 13
Shellfish: 28; 38; 39; 40; 43; 46
Shellfish allergy: 39
Shipping forecast: 95;
Shrimps: 18; 39; 43; 44; 45; 48; 50
Significant Wave Height (SWH): 101;
    102
Siphonophora: 34
SOFAR channel: 77
Sound: 76; 77; 102
Sounding: 79;
Spring tides: 106;
Squid: 33; 40; 41; 42; 47; 50; 58; 63;
    70; 71
Straits of Gibraltar: 3; 110
Stratosphere: 98
Subduction zones: 16; 22
Suez Canal: 3
Sunlight (see 'light ')
Suwa Maru No 58 (ill-fated Japanese
    fishing vessel): 102

Tectonic plates (see 'plate tectonics'):
Temperature: 7; 9; 18; 21; 28; 31; 32;
    48; 49; 98; 112
Territorial waters/sea: 118
Tethys: 3
Thermocline: 28;
Thermohaline circulation: 21
Tides: 104-111
    biggest: 109
    causes: 105
    Mediterranean: 110
    spring and neaps: 106
Tidal waves (see also Tsunamis): 111
Tonnage: 88
Trenches: 15; 16; 17; 22; 24; 27; 29; 43
Trent, River (tidal bore): 111
Trieste (bathyscaphe): 115
Tropopause: 98
Troposphere: 98
Tsunamis: 103
Tuna: 49; 52; 53; 61
Tuns (see tonnage)
Turbidite: 103
Turtles: 34; 35; 63

UNCLOS (United Nations Conference
    on the Law of the Sea): 118
Ungava Bay, Canada (record tide
    claimant): 109
Upwelling: 28

Volcanoes: 5; 17

Walsh, Ltnt Don, USN: 115
Waves: 93; 99; 100; 101; 102; 103;
    104; 111
Weather Reporter (ocean weather
    ship): 101
Whales general: 26; 32; 41; 42; 43;
    47; 50; 55; 58; 62-77
Whalebone: 63; 69; 71; 74; 75; 76; 77
Whale song: 77
Winds: 5; 19; 20; 21; 34; 93; 94; 95;
    97; 98; 101; 109
    - trade: 97; 98

Zooplankton: 15; 26; 32

blue and purple areas represent very low chlorophyll concentrations, while yellow and orange represent very high concentrations; the light blue and green areas are intermediate between these two extremes. Note that the very lowest concentrations (that is the poorest oceans) are the central sub-tropical areas, particularly the central southern Pacific. The richest areas are at high latitudes in both hemispheres and in very specific shallow areas bordering the land masses. For more details see Q28. (Reproduced courtesy of the NASA Modis-Aqua project).